Second edition published in 2022 by The Harvard Common Press, an imprint of The Quarto Group.
First Published in 2017 by Voyageur Press, an imprint of The Quarto Group,
100 Cummings Center, Suite 265-D, Beverly, MA 01915, USA.
T (978) 282-9590 F (978) 283-2742 Quarto.com

The Harvard Common Press titles are also available at discount for retail, wholesale, promotional, and bulk purchase. For details, contact the Special Sales Manager by email at specialsales@quarto.com or by mail at The Quarto Group, Attn: Special Sales Manager, 100 Cummings Center, Suite 265-D, Beverly, MA 01915, USA.

26 25 24 23 22   1 2 3 4 5

ISBN: 978-0-7603-7433-7

Digital edition published in 2022
eISBN: 978-0-7603-7434-4

Originally found under the following Library of Congress Cataloging-in-Publication Data
Title: The Brew Your Own big book of homebrewing : all-grain and extract
brewing, kegging, 50+ craft beer recipes, tips and tricks from the pros /
editors of Brew Your Own.
Other titles: Big book of homebrewing | Brew your own.
Description: Minneapolis : Voyageur Press, 2017.
Identifiers: LCCN 2016033863 | ISBN 9780760350461 (paperback)
Subjects: LCSH: Brewing--Amateurs' manuals. | BISAC: COOKING / Beverages /
Beer. | COOKING / Reference. | COOKING / General.
Classification: LCC TP570 .B8233 2017 | DDC 663/.3--dc23
LC record available at https://lccn.loc.gov/2016033863

Cover Design: Tanya R Jacobson
Cover Image: Shrimp & Grisettes
Page Design: Amelia LeBarron
Photography: (c) Brew Your Own, pages 58, 60, 63 (bottom), 64, 65 (top), 79, and 81; Charlie Parker, pages 1–156 (unless otherwise noted); Eric Gaddy, pages 160, 165, 176, 178, 192, 198, 212, 221, 225; Shrimp & Grisettes, pages 163, 166, 171, 173, 181, 187, 189, 195, 206, 208, 218, 227

Illustrations: Chris Champine, (c) Brew Your Own, pages 96–97

Printed in China

BIG BOOK OF HOMEBREWING

ALL-GRAIN <u>and</u> EXTRACT BREWING
KEGGING ✶ 50+ CRAFT BEER RECIPES
TIPS <u>and</u> TRICKS FROM <u>the</u> PROS

An UPDATED EDITION with
NEW TECHNIQUES, DOZENS
of NEW RECIPES, and MORE!

CONTENTS

INTRODUCTION

Every homebrewer has a story about when they first came to the realization that they could make their own beer at home. It may have been a friend or relative who shared a bottle of their own concoction; a beer that made them say, "I want to make this" that led to a conversation at a festival; or words on a page that sparked their imagination. And there are those who got into brewing because they desperately wanted to enjoy a style they couldn't find commercially. Others happen upon it out of curiosity and get wrapped up in the science behind the enzymatic conversion of the mash and the transformation taking place in their first fermentation. And, of course, there are those who just wanted to save some money (we know how that story ends).

My mind kept going back to the stories I've been told by homebrewers over the years as I contemplated how to start this book off. The more I thought about it, the clearer it became that it is the stories that make us homebrewers such a tight-knit community. Stories are entwined in everything we do, what we have learned through our successes and shortcomings, and what we pass along to those who come after us. No matter why they started, one aspect that keeps so many brewing after that first batch is that homebrewing is a social hobby that brings people together. Brewing is often the most fun when done with friends, but even for those who prefer to brew by themselves there is a sense of belonging to a group of hobbyists that spreads over the world. People connect and share what they know through homebrew club meetings, online message boards, and in text as our authors have done for over a quarter century. And after the brew day is done, final gravity is reached, and carbonation is achieved, people are once again brought together to share the latest brews at social gatherings while new stories are created.

In addition to the expansion of the hobby, this kinship has driven innovation. We are a creative bunch and pushing style limits, trying new ingredients or methods, and coming up with solutions to problems (including problems that may have never existed in the first place) is naturally a part of that. This constant tinkering has led to innovations that have changed what we think of as beer in modern times. Sure, brewing has been around for thousands of years, but much of what we drink today would likely never have been recognized as "beer" a couple of centuries ago. Heck, some of it may not have even been recognized as "beer" a couple of decades ago.

Brew Your Own magazine has been in the fortunate position of documenting, sharing, and expanding on these stories since our first issue was published in 1995. It was a demand for reliable, scientific-based, and proven information delivered in an enjoyable and nonintimidating style that led to the magazine's inception, and since then we have prided ourselves in being a leader in the homebrewing community. In each issue we strive to share information that every homebrewer can use, whether they are just opening their first bag of dried malt extract or mashing in batch #500. And that's precisely the goal of this newly updated book as well. With more than twenty-five years of experience, we are putting the best, most timely information in one place for you. Whether you are looking to learn the basics of dry hopping or the finer points of how water profiles impact the outcome of your beer, we believe this book will have something for everyone. Just like the first edition of *The Brew Your Own Big Book of Homebrewing* published in 2017, we'll lay out the different techniques to brew extract or all-grain batches, explore the wide world of brewing ingredients, and dive into the science of what's really happening along the way with a fun and easy-to-understand approach. Most of that hasn't changed in the past five years.

That doesn't mean the hobby is in the same place it was when we released the first edition, however. Brewing, and beer in general, has continued to evolve at a rapid pace in recent years. Hazy IPAs are even more popular now than ever before; however, they are seeing competition on store shelves from newer beer styles and beverages such as hard seltzer, pastry beers, heavily fruited "smoothie" sours, and beers fermented with kveik yeast, all of which we will explore in greater depth in this second edition. Among other updates, we have added a section on brewing kettle-soured beers, which has become a much more popular method to quickly sour beers since the first edition, and we've expanded the information on brew-in-a-bag, which is now even more mainstream with the rising popularity of the all-in-one automated brewing systems. These are just a handful of the topics we felt needed to be addressed in greater detail in this new second edition. We have been taking notice of these changes and itching to update these pages with the newest information available to help you brew the beer styles you want to drink, using the methods that best suit your lifestyle. In addition to laying out the ingredients and techniques to brew some of these latest styles, we've also included proven clone recipes provided by some of our favorite brewers, such as Drekker Brewing's Hyper Scream, a hazy IPA fermented with kveik yeast, Anderson Valley's kettle-soured Blood Orange Gose, and Southern Tier's Crème Brûlée pastry beer.

The story of homebrewing is still being told, and as long as grains are being malted and hobbyists are transferring wort to fermenters, we'll be here to help you brew the best beer possible. Who knows? That next recipe you have circulating in your head could be the next great story.

—Dawson Raspuzzi, editor

HOW TO HOMEBREW

TOP: Foaming sanitizers such as Star San are a popular choice among homebrewers.

BOTTOM: Sanitizing everything that touches your beer after the boil is especially important.

HOMEBREWING BASICS

Brewing is the process of making beer—a fermented, alcoholic beverage. You can think of beer as a beverage made from (essentially) four ingredients: malt, hops, yeast, and water. There are only a few basic steps to brewing as well: mashing (if you're brewing with malted grains), boiling, and fermenting. Homebrewing can be as simple or as geeky as you want it to be, but it's not difficult. Some brewers stick with basic ingredients and equipment, content to enjoy the process of a simple brew day. Others delve into numbers, experiments, building and expanding equipment, and competitions. Whether you want homebrewing to be simple or complicated is up to you, but before you make any decisions, this chapter will teach you the basics of brewing.

CLEANING AND SANITATION

Before you brew a single thing, we should discuss the fundamentals of cleaning and sanitizing. The first thing to note is that cleaning and sanitizing are best thought of as two distinct steps. The first step is to get the surfaces free of soils, scale, and stains. This isn't simply cosmetic; leftover soils have the potential to impart flavors that you don't want in your beer and create places for contaminants to hide during the next phase of the process. After cleaning, sanitizing ensures that biological agents that might spoil or damage your beer are (almost entirely) gone. In fact, "sanitized" is an EPA-regulated standard that requires the sanitizing agent to kill 99.999 percent of the pretreatment bacteria within 30 seconds. Once these two steps are completed, you can brew with confidence that whatever comes out of the other end of your process is the result of what you put into it.

GETTING STARTED

Cleaning is simple, right? Well, yes and no. On the one hand, simply scrubbing or wiping will clean. The addition of chemicals just makes that process more effective. Chemical cleaners use distinct agents to improve the effectiveness of cleaning. First, there are the parts that do the actual cleaning. Surfactants act as the "herding dogs" in the cleaning solution. They are designed to bond with soils, and as they do so, they gradually come between the soils and the surface. This creates a space between them and lifts the soils free, bit by bit, at the periphery—like slowly pulling off a Band-Aid. Once the soil is surrounded by surfactant molecules, it is no longer adhering to the surface that is being cleaned—hence, a chemically cleaned surface. Many modern brewery cleaners also take advantage of oxidizers, either instead of or in addition to surfactants. Oxidizers work by creating hydrogen peroxide, which pulls electrons from the soils, degrading them and breaking them down. Eventually, what's left of the soils no longer has the integrity to actually adhere to the surface anymore. But surfactants and oxidizers don't work alone; they need assistance to be effective in the liquid in which they're working, which is where builders, buffers, and chelating agents come in.

There's more than water in your water. Water is two parts hydrogen, one part oxygen, as most of us learned at some point in school. Mixed in with all that hydrogen and oxygen, though, are minerals and metals, which collectively make up our water. If your water is harboring sufficient levels of some additional elements (as hard water does), then your cleaning won't be as effective. Agents called builders and buffers can soften your water, regulate pH levels, and isolate or buffer the elements that might otherwise inhibit surfactants and oxidizers. Cleaners might also make use of chelating agents. Chelating agents are like bodyguards that allow the surfactants and oxidizers to work on what we want them to work on (the soils). Metal ions in the water can distract the cleaning agents and waste their efforts trying to clean the metal ions out of the water itself, instead of focusing on the soils. Chelating agents bind up those metal ions so that the surfactants and oxidizers ignore them and focus on the soils that we actually want cleaned and removed. Cleaning products may contain combinations of each of these kinds of ingredients, with different goals and effects, so there's no silver bullet in choosing a cleaning solution.

On to sanitizing! Sanitizing is, in many ways, much simpler. It requires only that you create an environment where bacteria can't live. This is accomplished by pushing the solution to the edges of the pH scale—either high acidity or high alkalinity are acceptable, and sanitizers are available that go in either direction. If the solution can reach the (cleaned) surface that you want sanitized, it will kill 99.999 percent of the bacteria present, and quickly. Sanitizing can also be accomplished by adjusting the temperature of the surface—though in this case, only one direction will get the job done (heat). Cold will just result in dormant bacteria, but heat kills. This process generally takes more time and isn't practical for all equipment, but it is an option (and it's why your brew kettle only needs to be cleaned, not sanitized).

CLEANERS

First and foremost, don't write off traditional cleaners. Most dish detergents can be used effectively on brewing equipment, with one very important caveat: avoid anything with added scents. You want basic, simple, not-lemony-fresh detergent. For your kettle, you might find that something such as Bar Keepers Friend is good for a seasonal scrubbing to help remove those burnt-in stains on the bottom of your pot. Traditional caustic cleaners afford a lot of punch for not much cost, but you'll need to be careful when using them. And for all of these, you'll want to rinse thoroughly, probably more than once, to ensure that you're not leaving behind chemicals that will endanger your beer (or yourself!), even if you will subsequently be sanitizing.

Most of the homebrew-specific cleaners we can now purchase, though, are of the noncaustic (and usually no-rinse) variety. They can roughly be grouped by their principle cleaning agent, whether oxidizers (those that break down and release soils through collection of electrons) or alkalines (those that use surfactants to loosen and remove soils from the surface).

One of the most common alkaline cleaners today is Five Star's PBW (Powdered Brewery Wash). You'll find the same basic cleaning and use parameters in Craftmeister Alkaline Brewery Wash, LD Carlson Easy Alkaline Powdered Cleanser, and other alkaline cleaners. These are generally noncaustic and safe to use without protective gear, but be sure to read the instructions on your product. One thing that we can definitely assume, though, is that they will make your equipment very slippery! Take extra precautions, especially when cleaning fragile and heavy items (such as a glass carboy). You'll also need to rinse anything cleaned with these prior to use. The upside, though, is that these are generally no-scrub cleaners. Even with particularly grimy carboys, kegs (be careful when using alkaline cleaners in closed equipment, though—pressure can build up), and tools, noncaustic alkaline cleaners should be up to the job on their own, without the intervention of scrubbing and brushes.

On the oxidizer side, you have a wide range of products to choose from. Probably the most common is Logic's OneStep, a no-rinse percarbonate (oxygen-based) cleaner. Similar products are available from most brewing chemical producers (Craftmeister Oxygen Brewery Wash, LD Carlson Easy Clean, Straight A Cleaner, and Fermenter's Favorite Oxygen Wash). At least one percarbonate cleaner does require rinsing (B-Brite), so be sure to read the use instructions on your product. However, most oxidizers do not require a rinse step. There is also some debate over whether they can be used as sanitizers as well, either in the same step or as an additional step (thanks to the release of hydrogen peroxide when mixed into a warm water solution), but this is still an open question. Hydrogen peroxide is a sanitizer, but it breaks down during the cleaning process, so its ability to sanitize may depend on the amount used, the amount of soil to be cleaned, and the contact time before it loses its molecular composition.

You may also find some specialty cleaners at your local homebrew shop that are designed for specific tasks. These include keg line cleaners such as National Chemical's BLC (Beer Line Cleaner), BLC Beverage System Cleaner, or Five Star LLC (Liquid Line Cleaner), and they are generally caustic acid or alkaline cleaners specifically designed for beer line cleaning. They will usually require a certain time/temperature circulation regimen to be effective and should be handled carefully and per the use instructions.

Whichever products you choose for cleaning and sanitizing, refer to and follow the manufacturer's instructions—not the Internet—to ensure that you get the best results.

SANITIZERS

Much like with cleaners, there are traditional options available in sanitizing. As mentioned earlier, one traditional no-rinse option is heat. Baking in an oven will kill (even to the level of sterilization) bacteria, but dry heat is only one option. Wet heat (steam) is another; many dishwashers now have a sanitize cycle, which creates steam of sufficient temperature to meet the sanitized standard. In practice, this requires that surfaces already be cleaned—and you should not include dish detergent as you normally would to clean dishes—but it can be a useful option if you already have the setting available.

In terms of dedicated brewing sanitizers, Star San by Five Star is a very popular product. It is an acid-based sanitizer, extremely fast-acting, and creates a foam that helps ensure that surfaces are coated. Five Star also offers a low-foam option (Sani-Clean) for applications that use a pump, and both are no-rinse when used below specific concentrations. There are also iodine-based sanitizers such as BTF Iodophor or IO Star, which are also generally no-rinse but can cause staining. In terms of spot- or surface-sanitizing, many brewers will keep a spray bottle of Star San, a dedicated surface sanitizer such as Alpet D2, or even cheap vodka on hand to provide a quick sanitizing option for specific tasks, such as the rim of a starter flask, the neck of a carboy, or the opening of a yeast pack. Sanitizers work (on clean surfaces) so fast that a quick shot is usually more than enough to allay fears of contamination.

PREVENTION

In terms of general advice on cleaning and sanitizing, you can do a few things to ensure that you're a conscientious homebrew janitor. First, clean in a timely manner. Don't put away dirty equipment. Don't allow kegs to sit empty, thinking, "Oh, I'll just clean that later." As soils age, they harden and become difficult to remove and can also result in staining. Second, when cleaning, consider the materials you're working with. Plastics scratch easily and can provide a safe haven for bacteria to hide from sanitizing agents. Some metals will react adversely to certain cleaning or sanitizing agents. Tailor your cleaning regimen to your equipment for the best results. Third, always use the prescribed amount/concentration of any cleaner or sanitizer. This is a case where more is almost never better, except within specific ranges. Manufacturers test their products in specific concentrations and construct them in specific formulations to accomplish the job they claim to do, and in many cases it can be dangerous to deviate from those concentrations (and usually it results in no better a cleaning/sanitizing job). Last, consider the effect of time. Not contact time (well, be aware of that, too—and again, follow the product's instructions), but brewing time. How many batches have you put through your equipment? Because the more you use it, the greater the risk of ineffective cleaning or sanitizing.

No matter how rigorous you are, you're still probably not achieving 100 percent efficiency in your cleaning and sanitizing. That means that whatever small amount of soil or bacteria you're leaving behind is potentially building up and accumulating over time. Maybe you don't have an issue for your first five batches . . . or ten . . . or fifty, but eventually you may find that you're getting off-flavors or contamination effects that you've never seen. If so, it isn't the end of the world. A lot of equipment can be replaced cheaply and easily, so resetting is an option. If you have stubborn soils that won't come off, or if you have a few too many scratches in your bottling bucket thanks to stirring in your priming sugar, you can always start over with a fresh piece of equipment. You can also implement more aggressive cleaning and sanitizing procedures that you go to only periodically—maybe you're fine with your standard cleaner for everyday use, but twice a year you go with a professional-grade caustic cleaner.

Your homebrewing equipment can be as simple or ambitious as you'd like.

EQUIPMENT BASICS

Step into any well-stocked homebrew shop and prepare yourself to be dazzled by all the stainless steel and gadgets. But while all the bells and whistles are fun, a basic extract brew day requires only a minimal amount of items to get you from start to finish. And if it's an all-grain setup you want, there are only a few extra items to add to the list. Once you have put together your homebrewery, you can add as many extras as you want as you move forward with the hobby, or you can (like many experienced homebrewers) just keep it simple—it's up to you!

EXTRACT EQUIPMENT

Despite the vast array of homebrewing equipment you can find for sale at homebrew retailers, making a basic batch of beer with malt extract really only requires a small collection of items, aside from the actual beer ingredients and standard cleaning and sanitizing agents. Many homebrew retailers even have starter kits already assembled, which makes shopping easy.

Here is what you need to brew with extract:

- A large brewpot: For classic extract brewing where you only boil about half of the wort, a 5-gallon (19 L) kettle will suffice.
- A metal or plastic spoon: This is for stirring the extract and ingredients into the wort. Metal and plastic are better than wood as they are easily sanitized.
- A fermenter: Most brewers make 5-gallon (19 L) batches, even for extract brewing. Make sure to choose a fermenter that is slightly larger than your batch size. Plastic brew buckets and glass or plastic carboys are easily sourced in 6-gallon (23 L) sizes. If you choose a carboy as a fermenter, it's a good idea to also buy a funnel for easier transfers from the brewpot.
- A plastic bucket for cleaning and sanitizing: 6 gallons (23 L) is a good size for this as well.
- An airlock plus a rubber stopper/bung
- A hydrometer and hydrometer jar: For taking specific gravity measurements before and after fermentation, you'll need this specialized tool.
- A racking cane, transfer tubing, and bottling wand: The racking cane and tubing are used for transferring your wort and beer from vessel to vessel. An auto-siphon model of racking cane (pictured on page 13) is inexpensive and extremely helpful. The bottling wand is an inexpensive addition and will make bottling day much easier.
- Bottles and bottle caps: If you're bottling 5 gallons (19 L), you'll need fifty-five 12-ounce bottles, or thirty 22-ounce "bomber" bottles.
- A bottle capper

If you want to get a little more advanced (with steeping grains, small mashes, and/or full-volume boiling), you should add these to your equipment list:

- A larger kettle equipped with a ball valve (pictured on page 13): The valve allows for easier transfers from the kettle to the fermenter, especially helpful at larger volumes. If you want to boil full-size (5-gallon/19 L) batches, go for something larger than your batch size to allow room for preventing boilovers. An 8-gallon (30 L) model should do the job, though a 10-gallon (38 L) kettle will be more useful if you make the jump to all-grain down the road.
- Reusable mesh steeping bags (for grains and hops)
- A wort chiller: This will even make partial-volume boil extract brew days shorter, and buying one is a great way to get ready for making the jump to all-grain brewing as well.
- A thermometer: You'll need an accurate thermometer for measuring the temperature of the water for steeping grains.

ALL-GRAIN EQUIPMENT

Traditionally, all-grain homebrew setups included three vessels compared with extract brewing's single vessel. However, in recent years many homebrewers have transitioned to all-grain with less equipment, thanks to a simplified brew-in-a-bag (BIAB) setup (see page 37), which requires just a single vessel.

For a traditional three-vessel system, the first vessel is used to heat all the water for your brewing session. As brewing water is sometimes called brewing liquor, the name of this vessel is the hot liquor tank, or HLT. Second, you'll need a vessel to hold the grains for both mashing (soaking the crushed grains) and lautering (separating the wort from the spent grains). This is called a mash/lauter tun. (In commercial brewing, these are often separate vessels, but homebrewers typically use just one vessel for both purposes.) This vessel needs to have a false bottom or some sort of manifold installed to let the wort flow from the vessel while retaining the spent grains. You will also need a large paddle or spoon to stir the mash (appropriately called a mash paddle). Lastly, you'll need a vessel to boil the wort in, called the kettle. Whether you use stainless-steel pots for all vessels is a personal choice. Stainless steel is more expensive but also more durable—and it gives you the most options as far as which vessels can be heated directly.

Two of the most common setups are as follows:
- Three 10-gallon (38 L) pots (usually stainless steel): You'll use one for the HLT, one for the mash tun, and one for the boil kettle. For these vessels, it's optional to have ball valves and fast-read thermometers installed for easy transfer and temperature reading. However, we highly recommend upgrading with these extras to make the brew day easier. Most brewers at least opt for the ball valves.
- One 10-gallon (38 L) stainless-steel pot and two 10-gallon (38 L) insulated coolers: In this setup, the pot is the boil kettle, one cooler is the HLT, and the other cooler is the mash tun. As mentioned with the three-pot system, ball valves are very useful (and often quite inexpensive to install in coolers). A thermometer on your brew kettle is a nice option too, although you can also use either a hand-held or a floating thermometer, which you'll need for your mash tun as well if you don't install a fast-read model on the vessel yourself.

No matter which way you go, if you haven't purchased one already, you'll need a wort chiller. Most homebrewers go with an immersion chiller, while some opt for a counterflow or plate chiller:
- An immersion chiller for 5-gallon (19 L) batches is typically made from about 15 to 30 feet (4.5 to 9 m) of copper tubing (although smaller chillers are available for brewing small batches/ kettles). A hose hooks up to both ends so that you can send cool water in one end and out the other. The cool coil chills your wort in bulk. As you chill, the wort eventually falls through a temperature range (from 160 to 120°F/71 to 49°C) in which wort contaminants are no longer killed by the heat and can in fact grow quickly in the warm environment. You should cover your kettle when the wort is passing through this temperature range to minimize the amount of airborne contamination.
- If you are using a counterflow or plate chiller, the surface of your wort in your kettle is always going to be near boiling, so you don't need to cover it as you chill. This type of chiller also forces you to decide what to do with the cold break particles (see page 35), as they are not left behind in the kettle as with an immersion chiller. However, using a counterflow or plate chiller allows late-addition hops, and especially the hops added at knockout, more contact time with near-boiling hot wort.

TIPS FROM *BYO*

BREWING SAFETY

Homebrewing is dangerous (well . . . sort of). Saying that homebrewing is dangerous is like saying that cooking soup is dangerous. If the pot was to fall off the stove and onto your feet, leaving you with second-degree burns, that would not be a pleasant thing. This outcome is not very likely unless you are being careless and inattentive, but nonetheless it is somewhat dangerous.

If and when you switch to all-grain brewing, you will find all-grain brewing requires some new equipment–and that once you go all-grain or take brewing outside, the safety level changes. Here are the things you need to watch out for in your homebrewery and some essential safety gear you should have easily accessible.

Venting propane: Propane burners release carbon monoxide and require ventilation. Brewing in garages with low or no ventilation can lead to carbon monoxide poisoning.

Cleaning quickly after spills: A small spill can cause a big slip and fall.

Bleeding pressure from kegs before opening lids: When you first start kegging, you may not fully understand everything about CO_2 pressure. Kegs under pressure should be bled of built-up gasses inside before attempting to open the lid. Failing to bleed off the pressure may cause the lid to fly off with great force directly into your face. People have died this way–take it seriously.

Testing the gravity of beer before bottling: Bottles will explode if you cap the beer before it has been fully fermented. Exploding bottles can cause a huge mess and are considered similar to grenade shrapnel. Brewers call these bottle bombs. Store bottles of homebrew inside a cardboard box to soften the blow in case something goes wrong.

Climbing ladders in gravity setups: Don't do this if you can help it. This is a quick way to fall, burn, or scald yourself.

Homebrewing in open-toed shoes: It's tempting to brew in flip-flops, but all kinds of things can spill or drop on your feet. Burns and smashed toes are not supposed to be part of the brew day.

Checking for loose hose connection clamps during hot liquid transfers: It's easy to forget to check your hose clamps, but sometimes they become loose for whatever reason. Firing up a pressurized pump can blow a hose off the pump, sending hot water and wort flying.

Moving full glass carboys: Glass carboy necks can snap off and shatter, resulting in lacerations. Never carry a full carboy by just the carboy handle. Products such as the Brew Hauler or Bucket Sling can help move a full glass carboy.

Moving to all-grain generally will require a change of heat source as well. With the larger volumes of water and larger vessels, stovetop brewing can lead to very long brew days (or worse, dangerous setups). Most brewers opt to move the brewing area outside and fire with a propane burner, as you've likely seen for turkey fryers. Lots of brewers also brew with all-electric systems or with induction burners. There are many choices when it comes to all-grain equipment—they will all depend on your individual needs. Keep in mind that great homebrew is made on a wide variety of brewing setups.

The final consideration is how your hot water and wort will be transferred from vessel to vessel. Some all-grain brewers use gravity to move their wort, which, for example, is the method employed in three-tier stands. Others opt to use pumps to move their wort from vessel to vessel. Choose what works best for your budget and setup.

BREWING WITH EXTRACT

In this section, we'll discuss making wort (unfermented beer) using malt extract accompanied by some steeped specialty grains. You can think of this as comprising three steps: making sweet wort, making hopped wort, and chilling the hopped wort. Sweet wort is the thick, sugary, not-yet-boiled solution made (in this case) from malt extract and steeped grains. Boiling the sweet wort and adding hops produces hopped wort, and chilling that solution yields wort that is ready to be fermented.

When making wort from malt extract, the ingredients aren't manipulated extensively—essentially, you dissolve the malt extract and boil it for a while. Along the way you also steep the specialty grains and add the hops to the boiling wort. Given that the process isn't that complex, your main goal is to get the most from your ingredients. In order to do this, it pays to understand their composition.

MALT EXTRACT

Malt extract is condensed wort in the case of liquid malt extract (LME), or dried wort in the case of dried malt extract (DME). Wort is mostly water, with the next most abundant component being sugars. Of the sugars, maltose is the most abundant. The minor components of wort (by weight) include proteins, amino acids, lipids, and all the various molecules that give wort its distinctive flavor, aroma, and mouthfeel. The flavor of wort comes largely from the sweetness from the carbohydrates in the malted grains, the "malty" (bread-like, toasty) flavors from the husk of these grains, and the bitterness and flavors from the hops. The practical upshot of this is that all you really need to do with malt extract is dissolve it in hot water to reconstitute the original wort. You do not actually need to boil it as you would wort made from malted grains—though nowadays most homebrewers buy unhopped malt extract and boil it as they would an all-grain batch for their hop additions.

Keep in mind that malt extract is a food product made from grains. Just as bread will eventually go stale, so will malt extract. Stale malt extract can be detected by its flavor and aroma, as well as by the fact that it becomes progressively darker as it stales. LME, because of its higher water content, goes stale faster than DME. Stored properly (in a cool location), LME will remain fresh for a couple months. Stored properly, cool and sealed away from moisture, DME may remain fresh for up to 8 months. Always strive to use the freshest possible malt extract when you brew.

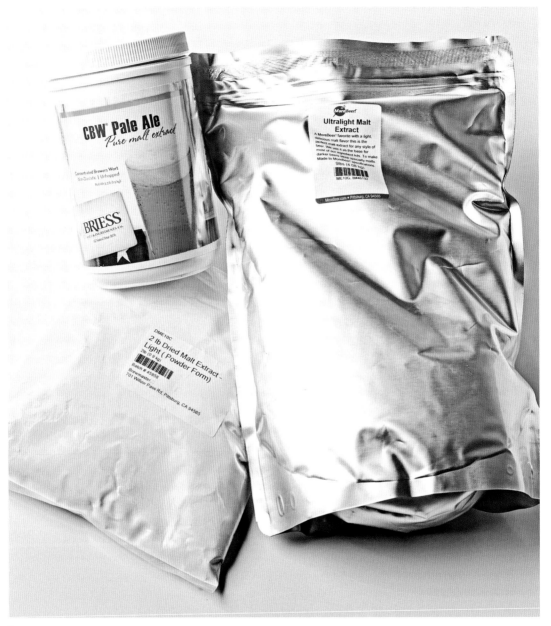

Malt extract comes in both liquid and dried forms.

MALTED GRAINS

Most extract recipes call for steeping some malted grains (usually specialty grains) to add to the flavor from the malt extract (usually pale, unhopped malt extract). As with malt extract, your steeping grains should be fresh. On brew day, your grains will also need to be milled (crushed). Most homebrew shops will mill the grains sold to extract brewers because most extract brewers do not own a grain mill. Unmilled grains will stay fresh for about 8 months, while milled grains will remain fresh for, at best, a week or so. So, if you get your grains milled at your homebrew shop, use them as soon as possible, ideally within a few days. Store them in a cool, dry place until your brew day.

Specialty grains contain sugars in their interior, which add to the original gravity of the beer you are making. More important, they impart flavors that are not present in the unhopped pale malt extract. These flavors come from the husk of the grains. Husks also contain a class of compounds you do not want in excess in your finished beer: tannins. These molecules cause a puckering astringency to your beer.

HOPS

Homebrewers mostly use either pellet hops or whole hops. Whatever form of hops you use, it is imperative that these hops be in good shape in order to get the best hop character in your beer. Hops should be green and smell fresh. Hops that have been stored poorly will turn brown and smell cheesy.

Hops should be stored frozen or refrigerated, preferably in packaging that blocks light and is flushed with an inert gas (such as nitrogen). Over time, even properly stored hops will decline in bitterness. The alpha acid rating can decline as much as 50 percent per year. However, the hop's aroma may remain appealing.

RECIPE CONSIDERATIONS

You can make high-quality homebrew from malt extract if you can find an appropriate malt extract for the beer you are making. Although there are malt extracts made from a blend of base malts and specialty malts, designed to make a particular style of beer, most homebrewers base their beers on a light or pale malt extract and add the additional malt touches by steeping specialty grains. Light, extra-light, or pale malt extracts are available that are suitable as the base for most English-style or American-style ales, and Pilsner malt extracts can be used for many German-style or Belgian-style beers. Wheat malt extract is also available for wheat beers. And, increasingly, you can also find malt extract made from Munich malt, Maris Otter pale ale malt, Vienna malt, smoked malt, and others.

Although you can make great beer from extract, it pays for extract brewers to understand that wort made from malt extract is not the same as wort made entirely from mashed base malt. First, because the extract is heated, albeit gently, during the condensation process, pale or light malt extracts yield worts slightly darker than the wort it was condensed from. In addition, the fermentability of a wort made from malt extract is usually lower than a wort made from malted grains. These drawbacks, however, only become a problem if you are attempting to brew a very light-colored dry beer. And, there are work-arounds. For example, you can subtract some of the malt extract from your recipe and perform a small partial mash of pale or Pilsner malt. Or, more simply, you can swap some of the malt extract in your recipe—up to about 20 percent—with table sugar (sucrose) or corn sugar (glucose). Either of these methods will lighten the color of your beer and increase the fermentability of the wort.

If you want to test the color of your malt extract, dissolve 2 ounces of dried malt extract or 3 ounces of liquid malt extract in a pint of warm water (57 or 85 g/250 mL); this will make a wort of specific gravity 1.048. This will show you approximately the color of a 5 percent ABV beer made from that malt extract, assuming it does not pick up any additional color in a long boil.

MALT EXTRACT BREW DAY

Once you've got your fresh ingredients assembled, your task as a brewer is to convert them into chilled, hopped wort—getting the best from each ingredient and ensuring that this wort will provide a healthy environment for the brewer's yeast.

1. Steep the Grains

Before any extract is used, the first step is to crush and steep any grains in your homebrew recipe. The quality of the crush is not as important in steeping grains as it is in mashing. As long as the grains aren't ground into a powder or mostly whole, you should be fine. (Your local homebrew shop will also likely crush them for you.) Place the grains in a steeping bag so they fill no more than one-third of the volume of the bag. This will allow liquid to flow past the grains while they are being steeped. It's a good idea to swirl the bag full of grains in the water a few times while you steep, but you don't need to do much more than this to get the full flavor from them.

Recommendations for steeping specialty grains run the gamut. To get the best character from your specialty grains, you should focus on two things: temperature and steeping volume. Most often, when steeping specialty grains, you want to get the same character from them as they would have imparted in an all-grain recipe. (If the recipe you are brewing is a commercial clone or an extract version of an all-grain recipe, this is certainly true.) To get the same character, you'll want to treat them in roughly the same manner. Therefore, in all but a few special cases, you

TOP: Check that your water is in the correct temperature range before you add the grain.

MIDDLE: Steeping grain in a bag makes it easy to remove the grain when it's time to move on to the next step.

BOTTOM: Rinsing the grain with additional warm water in a colander helps extract everything you should from the grain; it's a small-scale version of sparging (see page 32).

should steep your specialty grains in the temperature range of a single-infusion mash—148 to 162°F (64 to 72°C). If the extract recipe is a conversion from an all-grain recipe, steep them at the same temperature as specified for the mash.

The amount of water you steep the grains in is also important. If all of the steeped grains in an extract recipe are specialty grains, you have a fairly wide range that they can be steeped in—from so thick that the water just barely covers the grains to quite thin (by all-grain standards)—around 3 quarts of water per pound of grain (6.3 L/kg). If your steeping grains contain grains that need to be mashed, as is often the case in extract recipes in *BYO*, keep the water-to-grain ratio between 1.25 quarts per pound (2.6 L/kg) and 2.5 quarts per pound (5.2 L/kg).

For convenience, many old recipes instructed homebrewers to put the grains in their brewpot, filled to whatever volume was to be boiled. The grain bag remained in the brewpot until the boil started, or just before. In cases in which the water-to-grain ratio was high, this could lead to the extraction of tannins. Instead, if you want to get a jump on heating the water in your brewpot while you steep your grains, we recommend steeping in a separate pot and adding the "grain tea" to your brewpot when the steep is done.

Note: When brewing with malt extract, you don't need to worry about the mineral content of your water. The malt extract will contain the minerals from the wort it was condensed from. Your best bet is to reconstitute the malt extract using very soft water, such as distilled or reverse osmosis water. If the grains you are steeping contain some base grains, you can add a pinch (less than ⅛ teaspoon for a typical 5-gallon [19 L] recipe) of either gypsum or calcium chloride if you are using very soft water. If you are trying to brew a very hoppy beer and are using soft water to dilute the extract, adding 1 to 2 teaspoons of gypsum (calcium sulfate) per 5 gallons (19 L) will accentuate the hops. For more on brewing water, see page 118.

When your steep is over, do not squeeze or twist the bag to wring out every last bit of liquid as that may encourage tannin extraction. Your best solution is to let the bag drip until the liquid almost stops and then rinse the grains with a small amount of water at around 170°F (77°C). You can use a colander or a hand-held kitchen strainer and hold it over the brewpot. Most modern *BYO* recipes call for rinsing the grains at this temperature with around half of the volume of steeping water. This strikes a good balance between getting all you want from the grains, but not approaching conditions that would favor excess tannin extraction.

2. Add the Extract and Boil

The method of using malt extract that we recommend is to add a portion of the malt extract late in the boil. After you've added the "grain tea" from the steeping grains to your brewpot, add the remaining water for your boil and heat to just off the boil. Turn off the heat or remove the pot from the burner and stir in enough malt extract that your specific gravity is about roughly the original gravity of your beer. Withhold the rest of the extract until near the end of the boil.

For example, if you were boiling 2.5 gallons (9.5 L) of wort for a 5-gallon (19 L) batch, add a little less than half of the extract at the beginning of the boil. This, plus the sugars from steeping the grains, will make your wort roughly working strength. Then, add the remaining extract for the final 5 to 15 minutes of the boil.

One nice thing about brewing extract beers is the convenience. In particular, you only need to boil a few gallons of wort to make 5 gallons (19 L) of beer. But this convenience comes at a price. The thicker your wort is, the more likely it is to darken during the boil, and your hop extraction

LEFT: Turn off the heat when adding malt extract to ensure it doesn't stick and scorch on the bottom of the pot.

RIGHT: After adding malt extract and returning to a boil, you may see a lot of foam. Monitor the heat and reduce it as necessary to prevent a boilover.

efficiency is lowered as well. As such, as an extract brewer, always boil the largest volume that your brewpot size, heat source, and ability to chill allow.

Note: *If you notice a scorched or off-flavor in your beer, you may not be dissolving the extract sufficiently, and small amounts of it end up scorching on the bottom of the kettle. Next time, take samples as you brew. If the taste is there as soon as you add the extract, the extract may not be fresh enough. If the taste develops only after the first or second extract addition, then you have likely scorched some extract at the bottom of the brew kettle.*

3. Add the Hops

Extensive information on how adding hops throughout the boil affects your beer can be found starting on page 87, but it is almost certain that when you're starting out, your recipe will guide your hop additions.

How hops are added in the boil is an aspect of brewing where opinions diverge. Brewers that use pellet hops sometimes add them directly to the boiling wort while others place them in a bag so they can be removed at the end of the boil. If you bag your pellet hops, you need to leave plenty of room for them to expand. Don't fill more than one-third of the bag with pellets. Even filled loosely, however, adding the hops in a bag may limit your hop utilization. On the other hand, if you simply add the hops to your kettle, you will have to find a way to deal with the hop debris when the wort is cooled and being transferred to the fermenter. As such, you may need to leave behind some wort to avoid transferring too much gunk to your fermenter (part of what is called *trub*). Longer settling times allow the hop debris and trub to compact more at the bottom of the kettle, and whirlpooling the wort will ensure that the material is deposited in a cone near the center of the kettle.

If you add pellet hops directly to your wort, they have a tendency to cling to the side of the brew kettle, just above the wort line, after they dissolve. Be sure to knock these back down into the wort as the boil proceeds. They aren't contributing anything to your wort while they are clinging to the side of the pot.

After you add hops, you'll immediately notice a change in aroma.

Brewers using whole hops need a way to strain them out at the end of the boil. Some spigotted brewpots have a built-in strainer designed to screen out pellet hops, and these are also sold as add-ons. If you bag the hops, they won't expand as pellet hops do, but you still need to give them room for wort to flow past them. Don't fill the bag more than half full.

4. Chill the Wort

Stovetop extract beers, made from boiling a thick wort and later diluting it to working strength, are frequently chilled without the use of a wort chiller. If you are boiling 3 gallons (11 L) or less, you can quickly and efficiently cool your wort in a sink or bathtub filled with cold water and ice or ice packs. In addition, if you chill your dilution water, this will lessen the amount the thick wort needs to be chilled. If you chill in a sink ice bath, cover your brewpot and let the pot sit for 5 minutes or until the cooling water is too hot to touch, then change the water and add more ice. Repeat this a few times. Swirl the pot each time you change water (or stir with a sanitized spoon). You can also drape the lid of the brewpot with wet paper towels to cool the pot by evaporative cooling. If you boil more than 3 gallons (11 L) of wort, see page 35 for a better chilling option.

Note: *If your dilution water is colder than your fermentation temperature, you do not need to chill your wort all the way down to your fermentation temperature.*

5. Top Up, Aerate, and Ferment

At this point, your wort is ready to head to the fermenter. Transfer it from your brewpot, and add enough cool water to bring it up to full volume (if you haven't already during the chilling phase). It's okay to splash the wort around when you transfer from the brewpot to the fermenter as this will aerate the wort, which the yeast will appreciate. You should also shake the fermenter well once the wort is transferred to

TIPS FROM *BYO*

PARTIAL MASH BREWING

When partial mashing, you are getting a portion of the fermentable sugars for the wort from a mix of base and specialty grains. It is anything but difficult and requires only a little more time and attention to detail compared with steeping grains. The only extra piece of equipment that may be required is a good thermometer. You may also consider adding a bigger brew kettle to hold the entire volume of wort, a strong nylon bag to hold the grains, a device to quickly chill your hot wort to pitching temperature, and an aquarium pump and aeration stone to add oxygen to the wort.

MASHING (PARTIALLY)

Mashing is a simple process, but one that is often made to seem overly complex in some homebrewing texts. The essence of mashing is simply soaking crushed grains in water. As the grains soak, the water dissolves the starch in the grains. Enzymes from the grain attack the starch and chop it up into its building blocks: sugars. Once the starch is fully converted, the sugars are rinsed from the spent grains.

As far as starch conversion goes, a partial mash works exactly like a full all-grain mash. However, since less grain is used in a partial mash, handling the soaking and rinsing of the grains is simpler and requires no special equipment beyond a mesh grain bag and a measuring cup. Performing a partial mash is very similar to steeping specialty grains. Gaining some experience with partial mashing often encourages brewers to go on to try making an all-grain beer.

WHY PARTIAL MASH?

So why perform a partial mash? That's a little harder to answer but lies in the type of grains a recipe may require when it was originally designed as an all-grain brew. If an all-grain recipe calls for a modest to high percentage of unmalted grains (such as flaked grains or torrified grains) or nonsteeping specialty grains (such as biscuit malt, aromatic malt, honey malt, etc.), or uses a base grain that does not have a malt extract equivalent (such as rauch malt, Vienna malt, mild malt, etc.), then it is a good candidate for making a partial mash.

As already mentioned, in mashing the starch in the center of malted grains is broken down into its constituent sugars. By adding in some base grains with

further aerate the wort. Now you can pitch your yeast. Always be sure to pitch enough healthy yeast to make sure you run a strong, vigorous fermentation. You should see signs of fermentation within 12 hours, but hopefully even sooner. For more on yeast pitching rates, turn to page 42.

If you're fermenting in a bucket, place the cover on tightly and then get the bung and airlock (filled with water or water with sanitizer) in place. If you're fermenting in a carboy, simply push the bung and water-filled airlock into the neck. Place the fermenter in an area that is temperature appropriate to the recipe you are brewing, and let the yeast do its thing. Let the beer ferment until you reach your final gravity, which can be 3 to 5 days for an ale and longer (a week or more) for the unmalted or specialty grains, we get the enzymatic power to convert the starch into sugar, which the unmalted or specialty malt would be incapable of doing on its own. Here at *BYO* we always try to simplify all-grain recipes that have been submitted to an extract format for homebrewers. Some all-grain recipes easily can be converted to extract only, and some require the use of steeping grains such as crystal or roasted malts; some require performing a partial mash, and some just really can't be converted at all.

PERFORMING A PARTIAL MASH

See page 37 for instructions on brew-in-a-bag (BIAB). Performing a partial mash will follow the same steps, but you will then combine your resulting wort with the extract called for in your recipe. Keep in mind that at some point between going partial mash and full-on BIAB you will also want to increase your brewpot size to about 8 gallons (30 L) and invest in an immersion chiller so you can conduct full-volume boils.

Partial mash brewing is a great way to get a taste of what all-grain brewing is all about.

a lager. Leave the beer in the primary fermenter for a few extra days to condition, however, to let the yeast clean up some of the by-products produced by active fermentation (such as diacetyl—an undesirable compound produced during fermentation that tastes like butter or butterscotch).

Conditioning can vary based on the beer style and recipe, but for an ale this is usually about a week. For more on fermentation, turn to page 41.

6. Bottle or Keg

You are now ready to move on to bottling or kegging. For these steps, turn to pages 50 and 54.

ALL-GRAIN BREWING

In this section, we'll discuss making wort from malted grains. In homebrewing terms, this would be described as making wort using all-grain methods. As with extract wort production, there are essentially three phases: making the sweet wort, making the hopped wort, and cooling this wort in preparation for fermentation. Unlike extract brewing, the production of sweet wort is more involved and time-consuming. Essentially, the process includes soaking the crushed, malted grains in hot water. This is called mashing. Then the liquid wort is separated from the grain solids, and usually the grains are rinsed (sparged) to ensure a reasonable yield of sugars. This is called lautering. There are a number of variables that the brewer can manipulate that influence the quality of wort and quantity of extract achieved. To best understand the process, it pays to review the relevant characteristics of malted grain before we get to the processes.

All-grain brewing opens up a world of new malts and new flavors.

MALTED GRAINS

The most widely utilized grain in brewing is barley, followed by wheat. Although small amounts of unmalted grains are occasionally used in brewing, almost all brewing grains are malted. Malting is a process that readies the interior of the grains for the mash and develops flavors in the husk. Basically, the seed grains are soaked in water until they sprout, then dried to stop any further growth. Then they are kilned (heated in an oven) to develop the bready, toasty flavors of malt—and, in the case of specialty malts, the more darkly roasted flavors of crystal malts and darkly roasted malts, such as chocolate and black malt. Unmalted (seed) grain is very hard; malted grain is soft enough that it can be chewed. And, since malt can be chewed, you can taste it on brew day to ensure that it is fresh and lacking any of the flavors associated with staleness.

From the brewer's perspective, malted grains contain a starchy interior and a flavorful outer husk. The goal in making wort is to convert as much of the starch into sugar as possible and extract the best flavor compounds from the husks without extracting tannins (husk components that lead to astringency in beer).

THE CRUSH

How finely you crush your malt affects your brewhouse efficiency and the ease with which you can lauter your grain bed. The more finely you crush, the higher your extract efficiency. However, it becomes difficult to collect wort the more finely your malt is crushed. In addition, excessively

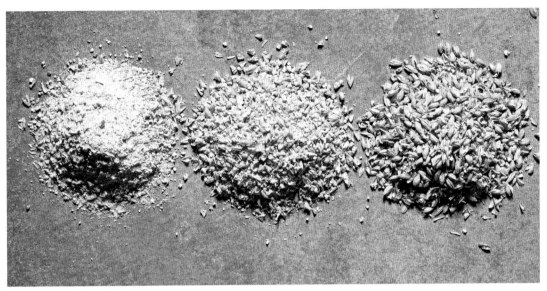

Make sure your crush is on target. From left to right in this photo, you can see a crush that's too fine, an ideal crush, and a crush that's too coarse.

finely crushed malt can yield more tannins when mashed and thus give your beer some astringency. Commercial breweries seek to have each husk broken into only two or three pieces and have the starch granules divided into large pieces, small pieces, and powder, with large and small pieces each constituting over one-third of the total.

If you crush your malt, experiment to find the right balance. Both your mill gap and the speed at which you crush will affect your crush. Faster-spinning rollers yield more finely crushed malt. The rollers on hand-cranked malt mills move slower than is optimum. However, when an electric drill powers the mill, the rollers spin much faster than is optimum. Either way, you will need to experiment with adjusting your mill gap to get the best crush.

Note: *If you are unsure about your crush—for example, if you are just trying out all-grain brewing—it is better to err on the side of undercrushed malt. If you crush your malt and all the kernels are broken open, that is enough. This will make lautering as easy as it can be (although you won't get stellar extract efficiency).*

MASH TEMPERATURE

Mashing has always been an extension of malting. In malting, the rock-hard barley seed is transformed into the relatively soft malted barley kernel. Along the way, the seed is modified in many other ways. In the mash, these modifications continue if the conditions are right. Long ago, brewers needed to begin mashing at lower temperatures—to take care of things such as gum degradation or protein modification—before raising the mash temperature to the saccharification range (148 to 162°F/64 to 72°C) to convert the starch to sugars.

These days, most malts are made so that all a brewer needs to do is employ a single-infusion mash—a mash with only one temperature rest. Modern malts are sometimes called well-modified malts, to indicate that almost all of the modifications that need to be accomplished have been achieved during the malting process. Undermodified malts, usually Pilsner malts, can be found with a little searching. They can be used if you want to do a multistep decoction mash or other multistep mash (see page 36).

In a single-infusion mash, the mash temperature is the primary way for a brewer to control the fermentability of the wort. If malts are mashed at the low end of the saccharification range (148 to 152°F/64 to 67°C), the resulting wort will be highly fermentable. The resulting beer will be dry compared with beers made from higher-temperature mashes. If you want to make an exceedingly dry beer (from wort with a very high degree of fermentability), you can add a rest—up to a couple hours—at 140 to 145°F/60 to 63°C before raising the temperature into the regular saccharification range. Alternatively, you can extend the mash time to 90 minutes if you are doing a single-infusion mash. Stir the mash as frequently as is feasible. (You can also substitute some highly fermentable ingredients such as sugar or honey for part of the grain bill.)

If malts are mashed at the high end of the saccharification range (156 to 162°F/69 to 72°C), the resulting wort will show a low degree of fermentability. The resulting beer will finish at a higher specific gravity and be more filling. If you wish to make a wort with a very low degree of fermentability, employ a short (20- to 30-minute) rest at the very top of the saccharification range (160 to 162°F/71 to 72°C), followed immediately by a mash out to 170°F (77°C). (Additionally, you can add some relatively unfermentable carbohydrates, such as lactose, to the recipe.)

You can pull a sample of your mash and test it with a drop of iodine to see if the conversion is complete. If the iodine is black, the conversion is not complete. If it fades to a light brown, the conversion is complete.

For most beers, you will be striving for an intermediate level of fermentability. The key to achieving this is to pick an appropriate temperature (152 to 156°F/67 to 69°C) and mash long enough to get a negative result on a starch iodine test (see above). Stir the mash and let the rest go to 30 to 45 minutes, if it hasn't gone on that long (to improve your efficiency a bit), then mash out by raising the temperature to 170°F (77°C).

MASH THICKNESS

The thickness of your mash also affects the fermentabilty of your wort. However, this effect is much less pronounced than that caused by the mash temperature. If a mash is exceedingly thick, the starch granules will not quickly or completely dissolve and the enzymes will not be able to diffuse through the liquid and reduce the starch to sugar. Likewise, in an excessively thin mash, the starch

  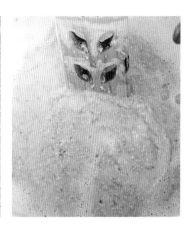

There's a wide range of acceptable mash thicknesses. Here you can see a thick mash, a normal mash, and a thin mash.

would dissolve, but the distances between starch molecules and enzymes would mean the mash would be slow to convert. In practice, there is a fairly wide window of mash thicknesses in between these extremes that works well. Anything in the 1 to 3 quarts per pound (2.1 to 6.3 L/kg) range will work. Homebrewers, especially those with limited space in their mash tuns, tend to favor fairly thick English-style mashes around 1.25 quarts per pound (2.6 L/kg). Brewers who make German-style lagers frequently favor thinner mashes, around 1.5 to 2 quarts per pound (3.1 to 4.2 L/kg) for dark beers and up to 2.5 quarts per pound (5.2 L/kg) for pale beers. If you are performing a step mash, thinner mashes are easier to stir when the mash is being heated.

WATER CHEMISTRY AND pH

When you dough in (stir the grains and brewing liquor together), the pH of the mash should settle into the 5.2 to 5.6 range (with the lower half of that range being preferable). Many times, this will happen without any intervention by the brewer if they are brewing a type of beer suitable for their water. The theory behind water chemistry and brewing is beyond the scope of this section, but a few points should be made. (For more on water, see page 120.)

When measuring mash pH, be aware that pH is temperature dependent. If you heat any solution, its pH will drop. As such, if you take a sample from your mash and cool it to room temperature before taking a pH measurement, you will need to subtract 0.35 from your reading to account for the rise in pH that accompanied the cooling of the sample. Cooling your sample is necessary for some pH meters and also helps prolong the life of the probe on many other models.

Arguably, the most important part of water treatment for brewers using municipal tap water is getting rid of the chlorine or chloramines used in water treatment. There are two possible ways to do this. One is to use a relatively large activated carbon filter. The filters that are housed under sinks should do the job while smaller filters (for example, the types that attach to a faucet) may not. The other is to add potassium metabisulfite, available at home winemaking shops in the form of Campden tablets. One Campden tablet stirred into 20 gallons (76 L) will neutralize any chlorine or chloramines. Because Campden tablets release sulfur dioxide (SO_2), you should let the water stand, loosely covered, for 24 hours to let the rotten egg smell dissipate. (At one tablet per 20 gallons/76 L, it's faint.)

ALL-GRAIN BREW DAY

When you have all of your ingredients assembled and your water chemistry dialed in, clear a few hours of your day and get ready to perform an all-grain brew.

1. Crush

Your homebrew store may have crushed your malt for you or you may be crushing the malt yourself. If you're crushing, see page 26 for more information on the crush size.

2. Mash In

Mashing in is the point at which you'll combine your brewing liquor (a.k.a. hot water) with the crushed grains. Your recipe will likely specify a water volume and/or a ratio for your mash thickness. We recommend using brewing calculators (see page 234) so that you can fine-tune your own process. For example, even if a recipe recommends a particular

Milling at home is optional. If you do your own milling, check that your crush is not too fine or too coarse (see page 27).

strike water temperature in addition to the mash temperature, the correct strike water temperature for your system will be affected by your ambient temperature and grain temperature on brew day—these are things a recipe writer just couldn't know. (The term *strike water* just refers to the water that you will use to mix with your crushed grains.)

For mashing in, you have options. You can add the brewing liquor to the crushed grains, or add grains to the brewing liquor. Of the two, adding your grains to your mash tun, then stirring in water, is the worst option. Stirring water into dry grains frequently leaves little malt balls, pockets of dried malt that can be difficult to break up. However, one small advantage to this method is you do not need to measure your brewing liquor—just keep stirring in water until you hit the correct mash thickness. This also allows you to take the mash temperature when the mash is almost mashed in but still very thick and—if needed—make small temperature adjustments to your brewing liquor in order to hit your target mash temperature.

If you fill your mash tun with hot brewing liquor, then stir in the grains, you will likely have no problems with malt balls. Plus, it's a lot easier to stir your mash. However, you need to measure the volume of your brewing liquor before you mash in and you do not have an opportunity to manipulate your mash temperature until you are completely mashed in. If you take good notes and are confident that by knowing your brewing liquor temperature and volume, you can hit your target mash temperature, stirring the grains into your brewing liquor is the quickest way to mash in.

LEFT: Before you mash, you may want to adjust your water chemistry (see page 29).

RIGHT: Stir as you mash in to prevent grain clumps.

3. Stir Occasionally and Monitor the Temperature

In most commercial breweries, the mash is stirred continuously. This evens out temperature variation throughout the mixture and increases brewhouse efficiency. On a homebrew scale, stirring will do both of these things, but frequently this comes at the price of losing heat to the environment. If you have a heatable mash tun, stirring the mash a few times (say every 10 minutes) and reestablishing the temperature by applying heat will likely increase your extract efficiency. If you are mashing in a cooler, at the high end of the saccharification range (with the aim of getting a less fermentable wort), stirring is not advised due to the inevitable heat loss.

During the mash, your only other task is to monitor the temperature. For single-infusion, you'll just need to keep it in the aforementioned range (either with direct heat or by adding additional hot water). For other mashes, such as step mashes, you may be following a mash schedule.

It's a good idea to stir the mash once or twice as time goes on so that the temperature remains even throughout.

4. Mash Out

After the mash, you have the option of performing a mash out—raising the temperature of the mash to 170°F (77°C) to make the grain bed easier to lauter and to greatly slow the action of the enzymes. If you are making wort with moderate to low fermentability, a mash out is highly recommended. If you can't heat your mash tun and don't have the room to stir in boiling water for a mash out, you can begin sparging with very hot water, 190 to 212°F (88 to 100°C), until the grain bed temperature reaches 170°F (77°C). At this point, cool the sparge water to hold the grain bed temperature at 170°F (77°C).

5. Vorlauf and Sparge

When the mash is complete, it's time to move on to lautering with the vorlauf and sparge. *Vorlauf* is also known as recirculating, and it is done to help the wort run clear (no big grain husk chunks) and to set up the grain bed as a filter. To do this, drain a few quarts of wort out through the valve in the mash tun and into a smaller container (a pitcher works really well). Now take the liquid you drained off and slowly pour it back into the top of the mash tun. Do this a few times until the wort starts to run mostly clear.

Now it is time to sparge. *Sparging* is the process of rinsing the grain bed with water after the mash to get as much of the converted sugars out of the grains as possible.

Continuous sparging is the most common method of sparging in commercial breweries, although in homebrewing, batch sparging may now be more popular. Often referred to as *fly sparging*, the continuous sparging method is a process of adding water to the mash at the top of the grain bed in a slow, continuous flow. You can do this very simply by slowly opening the valve on your mash tun and letting the wort start moving slowly into the kettle. At the same time, you'll add the sparge water into the top mash tun to try to match the rate the wort is flowing out of the kettle. You can do this manually by carefully pouring water into the top of the tun, or you can employ a sparging device to do the work for you.

Batch sparging is a simpler method than continuous sparging. It's done just like the name suggests: two or three batches of sparge water are poured into the mash tun, and the mash is stirred, rested, stirred again, and then run off.

Note: *Whenever sparging is described, using sparge water at 168 to 170°F (76 to 77°C) is almost always recommended. The idea is that tannins are extracted from the malt at an unacceptably high rate over this temperature. There are two problems with this idea. First, the temperature of*

You can either batch sparge or fly sparge with a sparging device (shown here).

TIPS FROM *BYO*

HOW MUCH WORT TO COLLECT

For any given weight of grain in your grist, there is an optimal volume of wort to collect from the standpoint of extract efficiency and wort quality. At this point, you have sparged the grain bed sufficiently to get a good extract efficiency, but not so extensively that you have extracted excess tannins. The exact grain weight-to-wort yield ratio depends on a few factors (including your crush and water chemistry), but expect to get a minimum of 6 gallons (23 L) of wort from every 10 pounds of grain. This works out to 0.6 gallons per pound (5.1 L/kg). If you are using continuous sparging, begin checking your final runnings when you have collected this amount of wort. Stop wort collection when the specific gravity dips below 1.008 or if the pH rises to above 5.8. In practice, many all-grain brewers simply collect enough wort to yield their target volume after the boil. For example, a brewer may always–regardless of how much grain a recipe calls for–collect 6.5 gallons (25 L) of wort and reduce it to 5 gallons (19 L) after a 90-minute boil. This actually works well for average-strength (1.040 to 1.057 SG) beers. However, for progressively higher-gravity beers, this means more and more sugars are left behind in the grain bed and extract efficiency will get progressively worse.

If you are interested in always getting the most from your grain bed, you will need to collect all the wort you can get from your grain bed, then boil this wort to reduce the volume. Of course, an all-grain homebrewer may be interested in brewing a high-gravity beer, but not conducting a long boil. In this case, collecting only the high-gravity first runnings and boiling for a (relatively) short time is the solution. Shorter boils not only save time but reduce the amount of color pickup in the boil. The only cost is the price of the extra malt. On the other side of the coin, when brewing a low-gravity beer, you may end up collecting too much wort if you collect your full preboil volume. In this case, you should collect all the wort you can, and then add water to make your full preboil volume. For example, let's say you usually get 75 percent extract efficiency and plan to brew 5 gallons (19 L) of dry stout with an OG of 1.038. A reasonable grain bill for this would weigh about 7.5 pounds (3.4 kg). If you used the ratio above as a guideline, you would collect 4.5 gallons (17 L) of wort and then begin checking the OG or pH (or both). Once you stop collecting, you would need to add water to yield enough volume to perform a 60- to 90-minute boil.

your sparge water is not what is important—it's the temperature of your grain bed. Given the small size of homebreweries, mash/lauter tuns can shed a lot of heat while you are collecting your wort. So, heat your sparge water to the point that your grain bed remains at 168 to 170°F (76 to 77°C) throughout wort collection, especially near the end. Second, tannin extraction is pH dependent. During most of the wort collection, your pH will be low enough that excessive tannin extraction will not occur, even at temperatures near boiling.

6. Boil and Chill

During the boil, the wort is sanitized, the hot break is formed, dimethyl sulfide (DMS) is volatilized, and the alpha acids in hops are isomerized. A lot goes on, but—if you've already decided on the total boil time and the timing of the hop additions—there is relatively little for a brewer to do. If your boil is less than rolling, a little stirring every 5 minutes or so might be helpful. Likewise, when the hot break forms at the beginning of the boil, removing the chunks of coagulated protein from it with a strainer can help improve the clarity and flavor of your beer (see page 35).

After the boil, you need to chill the wort to the point that you can pitch the yeast. The difference at this step versus the instructions for brewing with extract is that you will be chilling a full-volume boil and you will not be adding cool water to top up. It is important to chill your wort rapidly to make the cold break form (see page 35) and to slow the production of dimethyl sulfide (DMS), a compound that smells like cooked corn that is considered a beer fault. To quickly chill a full volume batch of wort, all-grain homebrewers typically use an immersion chiller, counterflow wort chiller, or plate chiller (see page 41 if you don't yet own this equipment).

7. Transfer and Ferment

Transferring a full-boil beer will typically be done through a spigot on your brewpot, though it is possible to transfer with a sanitized racking cane as well. Either way, the goal is a quick, sanitary transfer of the wort from the kettle to a ready and waiting sanitized fermenter.

Once the wort has been transferred, you'll want to aerate and pitch the appropriate quantity of yeast. See pages 41 and 42 for more on both of these crucial steps.

8. Bottle or Keg

Once fermentation is complete, you are ready to package your beer! Turn to pages 50 and 54 for more on bottling and kegging.

TOP: The goal is a nice rolling boil throughout. As with extract, watch out for boilovers.

BOTTOM: Chill the wort as quickly as possible with an immersion chiller or plate chiller.

TIPS FROM *BYO*

HOT BREAK AND COLD BREAK

Whether you brew with extract or with grains, if you properly brew your first batch of beer, you will experience two phases: hot break and cold break. Hot and cold breaks are stages when proteins in wort clump together so much that you can see floating particles. These particles are not only normal, but they are essential to brewing good beer. They are also visual signals that you have reached checkpoints in your brew day.

The first break you will experience during brewing is the *hot break*. Hot break occurs during the boil and precipitates when the wort reaches around 212°F (100°C). You can see the hot break forming when the foam (which is also coagulating proteins) stops forming and you start to see gelatinous-looking clumps floating in the pot. This happens because the proteins' structures change (or denature), much like a raw egg white becoming solid as it is heated. Depending on the amount of protein in your wort, this can take anywhere from 5 to 20 minutes. At the end of the boil, the hot break is allowed to fall to the bottom of the kettle as part of the trub, which is the material left behind when the wort is siphoned from the kettle to a fermenter.

Achieving a good hot break is important as it will encourage proteins to fall out of suspension. Too many of those proteins in your wort can cause the beer to become hazy, exhibit off-flavors, and can also lower the beer's long-term stability.

After the boil, wort should be rapidly chilled to yeast-pitching temperatures to prevent contamination and oxidation. This rapid chilling process also causes *cold break*, which is the stage when another group of proteins precipitate from the wort. These proteins stick together, as well as to other particles (like tannins) still in suspension in the beer, and fall to the bottom of the fermenter when they are allowed to settle. Cold break will look similar to the clumps created during hot break, and it starts to form at temperatures around 72°F (60°C).

Cold break is important for similar reasons that hot break is necessary. But also, cold break is important because it is your last chance to remove some of those proteins—anything left behind by not properly chilling your wort can create chill haze. Chill haze happens when remaining proteins bind together when the beer is cold; when the beer warms up, the bonds dissolve back into the beer. If the beer is heated and cooled too many times, the haze can become permanent.

Achieving a good hot or cold break is simple if you are patient, pay attention to what's going on in the brewpot, and keep your eyes open on the wort for signs of the breaks forming. Hot and cold break also signal that you have done a good job boiling or chilling and are the points when you can move on to the next step of the brew day.

To get the best possible hot break, be sure to bring your wort to a hot, rolling boil. If you can't get your beer up to rolling boiling temperatures, upgrade your kettle heat source to something that can boil a full batch. Many brewers also add fining agents such as Irish moss or Whirlfloc to the boil to help precipitate the proteins.

To get the best cold break, always rapidly chill your wort. This means cooling your wort from boiling temperatures (212°F/100°C) to yeast-pitching temperatures (70°F/21°C) in 30 minutes or less. Slow cooling does not affect the cold break-sensitive proteins, so if you experience difficulty rapidly chilling your wort with your existing equipment, or if you are not already using a wort chiller, you may need to find a more efficient chilling method.

MASH METHODS

Now that you're mashing, you may be interested in knowing about various mash types. A few mash options are more popular than the rest with today's brewers: single-infusion, step mashing, and decoction. As you get more advanced, you will hear about other mashes as well, including cereal mashing and sour mashing. Let's go over these different techniques.

SINGLE-INFUSION

In the single-infusion mash, a known quantity of water at a known temperature is combined with a known volume of grain at a known temperature. Through the use of temperature charts or formulas, this results in a very consistent method of obtaining repeatable mash temperatures that are accurate within a few degrees. By shifting variables such as the temperature or volume of the strike water (the water into which you mix your grains), the final temperature of the mash can be adjusted up or down.

STEP MASHING

Temperature program or step mashing differs from single-infusion mashing in that numerous temperature rests can be incorporated in the mash schedule. Resting the mash at a series of successively higher temperatures allows a wider variety of enzymes to work within their own optimal temperature ranges for a set amount of time. This results in more control for the brewer over the final characteristics of the sweet wort and the ensuing beer.

You can raise the temperature of the mash through the use of a mash tun to which heat can be applied or through repeated infusions of hot water to successively raise the mash to the appropriate rest temperatures.

- The beta-glucan rest occurs at 118°F (48°C) to reduce wort viscosity caused by gummy beta-glucans. This is especially important when using flaked barley, oats, and rye, as well as lightly modified malts.
- The beta-amylase rest occurs at 135 to 140°F (57 to 60°C) to increase wort fermentability, while the alpha-amylase rest takes place at 158°F (70°C) to "convert" the mash.
- Two rests no longer used by most commercial brewers are the acid rest at 110°F (43°C), which was formerly used to adjust mash pH, and the protein rest at 122°F (50°C), which current research indicates has no value.

DECOCTION

Decoction mashing is a form of step mashing in which the brewer removes a portion of the mash, usually about one-third of the thickest part, leaving behind as much of the liquid as possible. This thick decoction is placed into a separate kettle, where the temperature is raised until the decoction boils. Then the decoction is reintroduced into the main mash, thereby raising the temperature of the whole mash to the next step. This entire process can be repeated as many as three times to incorporate the same temperature rests discussed in step mashing.

The reason for doing this is that when the thick grain portion of the mash is boiled, a physical breakdown of the starches and proteins results, giving the enzymes present greater access to them. This results in a greater conversion rate (more starches are converted to fermentable sugars) than with other mash methods, particularly when low-modified or undermodified malts such as Pilsner and lager malts are used.

This style of mashing evolved on the European continent, where the malts available at the time were undermodified and lacking in the enzymatic power of their English counterparts. These days malts are much improved, and unless you are brewing to follow tradition, you probably don't need to perform a decoction mash.

CEREAL MASHING

Cereal mashing allows you to mash corn grits, rice, or unmalted wheat for certain beer styles. It also allows you to experiment with virtually any starchy food. One of the advantages of all-grain brewing is the ability to use ingredients that can't be used in extract brewing. Specifically, all-grain brewing allows brewers to use starchy grains or adjuncts that would cause haze (and instability) in an extract beer. Because grain-derived enzymes in the mash (alpha- and beta-amylase) degrade starch into simple sugars, starchy adjuncts can be added to an all-grain mash. In order to degrade starch in a mash, however, the starch needs to be accessible to the starch-degrading amylase enzymes. When you plan to utilize a grain or starchy adjunct using a cereal mash, you will need to plan whether you want to do a step mash or a single-infusion mash.

A traditional cereal mash, of the type used in making American Pilsners, is part of a step mash program. In this mash program, the barley is mashed in at a temperature below the starch conversion range. The cereal mash—which can be 40 or 50 percent of the size of the main mash—is boiled, then pumped into the main mash. The heat from the cereal mash raises the temperature of the main mash. As a homebrewer, however, you can also opt for a single-infusion mash along with your cereal mash. To do this, boil your cereal mash, but do not mash in the rest of your grains initially. When the cereal mash is ready, combine it with the crushed grains and hot water to mash in at your preferred temperature.

To do a cereal mash, combine your "cereal"—whether it's corn, rice, an unmalted grain, or other starchy food—with about 10 percent 6-row barley malt or 15 percent 2-row barley malt. The malt should be crushed and—if your cereal is another grain—crush that too. Slice, dice, or otherwise reduce the size of other starchy foods to small-enough pieces so that they will hydrate quickly. You can go higher on the barley percentage if you want, up to around 30 percent.

Add water and begin heating the cereal mash. Shoot for a thin gruel-like consistency. Some foods will take on water as they cook, so don't be afraid to add water as you go if the cereal mash gets too gooey. Bring the cereal mash to the high end of the starch conversion range, around 158°F (70°C), and hold for 5 minutes. The barley malt in the mix will convert any stray starches at this point, but the bulk of the starches will be converted in the main mash. (Even with starchy foods with a low gelatinization range, there is not enough enzymatic power in the cereal mash to fully convert it.)

After the 5-minute rest, bring the cereal mash to a boil. You will need to stir nearly constantly as it heats and boils to prevent scorching. Boil the mash for 30 minutes. When the cereal mash is done, stir it into your main mash. At this point, the starches in the cereal mash will be exposed to the amylase enzymes in the main mash and degraded. From here, you simply finish brewing as you normally would.

BREW-IN-A-BAG

Brew-in-a-bag or brew-in-a-basket (BIAB) is an easy and economical way for homebrewers to start all-grain brewing. What makes this method unique for all-grain brewing is that you can brew with one kettle instead of two kettles and a mash tun. With a single brew kettle and a large fabric filter (available at most homebrew supply stores), you have all that's needed to brew an all-grain beer. BIAB is also a great way to perform a partial mash that you can add to an extract brew (see page 24).

TIPS FROM *BYO*

TROUBLESHOOTING A STUCK SPARGE

Lautering is generally a very straightforward, uneventful process, but occasionally the dreaded stuck sparge will show up and ruin your otherwise pleasant brew day. A stuck sparge occurs when the flow of sweet wort from the mash-lauter tun slows to a trickle or, worse yet, completely stops. Luckily, the homebrewer can do a number of things to help prevent stuck sparges from ever occurring in the first place.

- The most important step in preventing a stuck sparge is milling your grain properly, as it is the husk of the grain that forms the actual filter bed in the tun. Milling too fine will encourage stickier sparges.

- Consider adding rice hulls to the mash at a rate of 0.25 to 0.5 pound (0.11 to 0.33 kg) for a typical 5-gallon (19 L) batch of beer. Rice hulls act as a filter aid by preventing gummy grain from forming clumps and impeding the flow of wort though the grain bed. Rice hulls are most effective when added to the milled grain before the mash is started, but they can also be stirred into the mash after it is complete if you've found yourself with an unexpected slow or stuck runoff.

- Consider a beta-glucanase rest, during which the temperature of the mash is held at 98° to 113°F (37° to 45°C) for 20 minutes before proceeding with the saccharification rest. This step will break down some of the gummy beta-glucans in the mash, allowing it to flow more easily.

- Perform a mash out as described on page 32. Also, if you're utilizing a fly sparge, it is important to maintain approximately 1 inch (2.5 cm) of liquid above the top of the grain bed at all times, as allowing the grain bed to run dry will cause it to compact, making it more difficult for the sparge water to flow through.

Stuck sparges are an inconvenience, but you can usually resolve them without too much trouble.

So-called all-in-one single-vessel automated brewing systems generally use the BIAB (often using a basket). These systems come in various shapes and sizes and have a diverse array of features depending on the model; however, the mash is often conducted in this same way.

There are some special considerations when it comes to BIAB mashes vs. traditional mash methods. The most important is your water. The high water-to-grist ratio can cause problems with the mash pH if you have high alkalinity water—and most people do. Look at your water report for Total Alkalinity as Calcium Carbonate. If this number is 100 ppm or greater, your alkalinity is high. If it is more than 150, then it is very high. If you have high or very high alkalinity water, a 50/50 dilution with distilled water and/or using some acidulated malt or brewing acid (such as lactic acid) to help neutralize the alkalinity and bring the mash pH down into the recommended range of 5.2–5.6 is recommended. Another option would be to use a more typical water-to-grist ratio of 1.5–2 quarts per pound (3–4 L/kg) to conduct the mash, and then add the rest of the water at the end. If adding water at the end of the mash, stir to get everything homogenous, let it rest a few minutes, and then drain as usual. The idea is that you are optimizing the pH conditions for the mash at the lower water-to-grist ratio and then adding more water later to get your total boil volume and gravity, without sparging.

The next consideration is extraction efficiency, which primarily depends on the water-to-grist ratio. Generally speaking, the efficiency of BIAB should be between 74–84 percent for beers with OGs between 1.040–1.075, lower OG having higher efficiency than high OG. One way to increase the efficiency of BIAB brewing is by increasing the water-to-grist ratio. Crushing the grist finer/ smaller is another way to increase efficiency by a few percent and because there is no lautering in BIAB there is no need to worry about a stuck mash. For this reason, many homebrewers will double crush their grains to maximize their efficiency. More good news in terms of efficiency—the BIAB water retention factor, 0.25 quart per pound (0.5 liter per kg) of grain, is typically half that of a standard mash.

1. Add Strike Water and Heat

Place your kettle on the heat source and add the volume of water required for your BIAB recipe or partial mash. Your volume of water matters, so be sure you've calculated properly. Try using any number of free BIAB calculators, such as www.simplebiabcalculator.com or www. biabcalculator.com. These calculators account for factors such as boil off, kettle diameter, grain absorption, trub loss, and so on to determine your strike water volume and target temperature.

Mashing in a large mesh bag means you need less equipment on brew day.

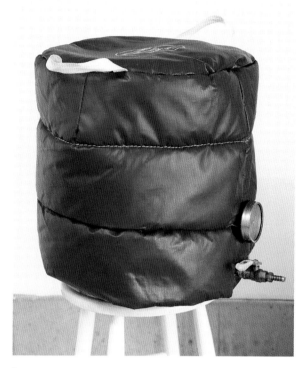

2. Insert the Bag

Once you reach your desired strike temperature, turn the burner off and insert the bag into the kettle if you haven't already done so. Secure the bag with a bungee cord or clip the bag to the edge of the kettle with binder clips or anything that allows the lid to be put on while preventing the bag from slipping into the kettle.

3. Add the Grain

Add the grains to the bag. Take a temperature reading by inserting a thermometer in four sections of the mash. If the temperature varies, stir a bit more and measure again.

4. Mash

Now cover the kettle and mash the grains for anywhere between 45 and 90 minutes. If you have the means, insulate the kettle to keep the temperature within 2 to 3 degrees of the initial mash temperature. A kettle jacket works great; if you're using a small kettle, you can preheat your oven to 150°F (66°C) and put the

You can buy special insulators that fit large pots–they help keep your mash temperature stable.

kettle inside. The majority of conversion takes place in the first 45 minutes of the mash, so even if a couple degrees are lost over that time, the starches are converted in the proper range.

5. Check the Temperature and Stir

After the mash is complete, remove the insulation and the lid. Stir the mash well and check the temperature and record it. Now push the bag aside and take a sample of the wort to record preboil gravity.

6. Lift the Bag and Drain

Since you're doing a partial mash or full-volume mash, there is no need to sparge. If your kettle is under 8 gallons (30 L), you can likely lift the bag with your hands. For anything larger than that, or if leverage is an issue, you'll need a pulley and hook to safely remove the bag. If you have a kettle thermometer installed in your kettle, spin the bag gently as you lift to ensure the bag is free. No matter how you remove the bag and in order to capture as much wort as possible, find a way to gently squeeze the wort out of the grain and into the boil kettle.

7. Boil and Beyond

With the bag and grain out of the way, fire up to boil. You'll now follow your recipe for the remaining ingredients/hop schedule and the rest of the process.

BIAB is not limited to single temperature rests. Many brewers will maximize the fermentability of their wort by doing both a beta and alpha amylase rest. If you are mashing in a kettle, simply raise the bag off the bottom and stir while you are heating from the stove or burner. If you are mashing in a cooler, then start the mash at a lower water-to-grist ratio, such as 1.25 or 1.5 quarts per pound (2.5–3 L/kg), and use infusions of hot or boiling water to raise the temperature to the next rest. Be sure to add the hot water slowly and stir during to reduce the stress on the enzymes. Decoction is another good way to conduct multiple temperature rests (these processes are further described on page 36).

FERMENTATION AND CONDITIONING

In this stage of brewing, the brewer must create the proper conditions that allow the brewer's yeast to transform the hopped wort into beer. After fermentation, brewers may also need to set the correct conditions for the yeast to clean up some molecules produced during fermentation (such as diacetyl).

NUTRITION AND AERATION

You'll want to start by considering yeast nutrition. As well as having enough oxygen—which we'll get to in a moment—yeast also need a healthy amount of nitrogen, vitamins, and minerals. For most all-malt beers, this is not a problem. If a wort is deficient in anything, it is likely zinc. Adding commercial yeast nutrients at a rate ranging from half to the full manufacturer's recommended rate during the boil will almost always solve the problem.

After the boil, once your wort is chilled, it is time to aerate to add the oxygen. If you have a counterflow or plate chiller with an aeration stone placed on the wort outflow side, you can aerate your wort as it flows into your fermenter. Or, as is more common, you can aerate the wort using an aeration stone and bottled oxygen or an aquarium pump. Frequently, brewers place a HEPA filter between the stone and the source of air or oxygen to ensure no microorganisms enter the wort through their aeration efforts.

Optimally, wort should be aerated so that it contains between 8 and 10 parts per million (ppm) dissolved oxygen. Unfortunately, homebrewers do not have an inexpensive way to measure oxygen levels in wort. Dissolved oxygen (DO) meters have come down significantly in price in recent years, but they are still more expensive than the grade of pH meters most homebrewers use. You can stipulate that the wort be aerated for a certain amount of time, but this doesn't account for the flow rate of the gas or the size of the holes in the aeration stone. Even if you could measure the volume of gas pumped into the wort, you have no easy way to determine how much has been retained. This depends on a few variables, including temperature. Gas dissolves more slowly the colder the wort is (although the capacity to hold gas goes up with lower temperatures) whereas swirling the fermenter while aerating increases gas diffusion.

Still, some basic guidelines can be given, and homebrewers can infer if they work or not by observing their fermentations. Generally, with a stainless-steel airstone, 1 to 2 minutes of oxygenation—during which a constant cloud of tiny bubbles is coming from the airstone—should be enough to aerate a batch of beer. Likewise, 5 to 10 minutes of air (for example, pushed by a fish tank aeration pump) should get you to the minimal required level. It is possible (albeit unlikely, if you are following normal aeration procedures) to overaerate a batch using oxygen, but not with air.

Homebrewers need to take care to monitor their fermentations early on to see if the yeast has received adequate aeration. Adequately aerated ales should start fermenting within 24 hours. Lagers should start fermenting within 36 hours. The amount of time until a fermentation starts also depends on the yeast strain, pitching rate, wort temperature, and level of wort nutrients, and therefore can vary quite a bit. Start times can be much sooner than the times given earlier, especially with healthy yeast at higher pitching rates. Incidentally, after aerating your wort, gas will begin diffusing out of solution and back into the atmosphere. For this reason, have your yeast ready to pitch immediately after aeration.

PITCHING RATES

You need to pitch an adequate amount of yeast to get your fermentation to start in a reasonable amount of time, proceed in an orderly fashion, and reach a reasonable final gravity (given the fermentability of your wort). A pitching rate of 1 million cells per milliliter per degree Plato is frequently cited as the standard rate for ales, although some sources give a lower rate. For a 5-gallon (19 L) batch of beer at 12°Plato (SG 1.048), this would be 228 billion cells. The optimal pitching rate for lagers is often given as twice this, although again lower rates can be found in the professional literature.

To accurately measure the amount of cells, you need a microscope, a special kind of slide called a hemacytometer (designed to count blood cells) and a vital stain (methylene blue). As most homebrewers do not have this equipment, most rely on pitching a given weight or volume of a yeast slurry, pitching yeast from a yeast starter of a given volume, or by pitching multiple packages of commercial yeast based on their cell counts.

For a 5-gallon (19 L) batch of moderate-strength ale, a long-standing rule of thumb has been to pitch a cup of yeast slurry. For homebrewers repitching yeast from the bottom of a fermentation bucket or carboy, this often works well because the density of yeast cells in the slurry immediately after fermentation is relatively low. This yeast sample will be liquid-like and colored with trub and hop debris that settled along with the yeast. If you harvest healthy yeast and let it settle overnight in your refrigerator, about one-third this volume (⅓ cup/80 mL) would be satisfactory. Yeast selected this way will be creamy to pasty in consistency. And, since the trub and hop debris will sediment in separate layers, it is relatively easy to use only yeast slurry, which will be off-white in color.

If you are making a yeast starter (see page 43), you can estimate the amount of cells you will raise from a given volume of starter wort. The density of yeast in a well-aerated yeast starter would vary depending on yeast strain and other variables, but 50 million cells/mL to 100 million cells/mL is not an unreasonable estimate. If you calculate the total number of cells you need to pitch, simply divide this number by the density of your yeast starter to yield the size of the yeast starter (in milliliters). Or, see the table on page 43 for starter sizes for three different pitching rates over various original gravities from 8° to 16°Plato. The website www.mrmalty.com also has a calculator that suggests a suitable yeast starter volume for a given volume and gravity of wort.

A rule of thumb *BYO* has used in the past is that, for moderate-strength ales, a 2-quart (2 L) yeast starter is optimal. Mr. Malty returns a value of half of this (for yeast starters initially aerated with oxygen), indicating that those calculations are based on slightly different assumptions. In reality, yeast density varies depending on yeast strain, aeration of the medium, nutrient availability, and other things. If you aren't counting your yeast, you are relying on assumptions you can't test. In practice, however, beer is fairly forgiving; if you raise a healthy yeast starter and are within the ballpark of the optimal pitching rate, your beer will likely be fine.

If you are using dried yeast, making a yeast starter may be counterproductive. Dried yeast has a high amount of glycogen stored in it, and making a yeast starter (especially if the starter is too small) may deplete that store of glycogen. When using dried yeast, some homebrewers like to rehydrate the yeast in water immediately before pitching; however, dry yeast manufacturers are now saying that yeast can be pitched directly into the wort to no ill effects.

Higher pitching rates generally lead to faster starts, quicker finishes, and higher attenuation. In addition, the amount of yeast character is lower in beers pitched at a high rate. If you are brewing a beer that benefits from some yeast character (esters, etc.), as is the case in most English and Belgian ales, pitching at a less than optimal rate can help accentuate the yeast by-products as these are mostly formed when the yeast are multiplying, as opposed to when they are at a roughly constant number and fermenting.

YEAST STARTER SIZE BY PITCHING RATE

Original gravity of beer (SG)	Ales (low optimal) 0.75 million cells/mL/SG	Ales (high optimal) 1.0 million cells/mL/SG	Lagers 1.5 million cells/mL/SG
1.032	1.2 qt. (1.1 L)	1.6 qt. (1.5 L)	2.4 qt. (2.3 L)
1.036	1.3 qt. (1.3 L)	1.8 qt. (1.7 L)	2.7 qt. (2.6 L)
1.040	1.5 qt. (1.4 L)	2 qt. (1.9 L)	3 qt. (2.8 L)
1.044	1.6 qt. (1.5 L)	2.2 qt. (2.1 L)	3.3 qt. (3.1 L)
1.048	1.8 qt. (1.7 L)	2.4 qt. (2.3 L)	3.6 qt. (3.4 L)
1.053	1.9 qt. (1.8 L)	2.6 qt. (2.5 L)	3.9 qt. (3.7 L)
1.057	2.1 qt (2 L)	2.8 qt. (2.6 L)	4.2 qt. (4.0 L)
1.061	2.2 qt. (2.1 L)	3 qt. (2.8 L)	4.5 qt. (4.3 L)
1.065	2.4 qt. (2.3 L)	3.2 qt. (3 L)	4.8 qt. (4.5 L)

PITCHING TEMPERATURE AND CONTROL

When you pitch your yeast, you should take care not to thermally shock them. In general, your yeast should be within 10°F (5°C) of your wort temperature. If you brew lagers and raise your yeast starter at room temperature, cool the starter solution in your fermentation chamber (which should be set to a couple degrees below your planned fermentation temperature).

Once the yeast have been pitched, the main goal of the brewer is to maintain the temperature of the fermentation to produce the best beer. Beer yeasts grow best at temperatures above that which produces a quality beer. Most ale yeasts produce the best beer at 65 to 72°F (18 to 22°C), and most lager yeasts work best at 50 to 55°F (10 to 13°C). In some Belgian ales, fermentation temperatures are allowed to climb much higher than in English ales (up to 85+°F/29+°C).

The most common way to maintain proper fermentation temperature at a homebrew scale is to place the fermenter (bucket, carboy, or stainless-steel fermenter) in an environment that is a few degrees colder than the planned fermentation temperature. Sometimes this simply means placing an ale in a cool spot in the basement. Other times it means placing the fermenter in a fermentation chamber made from a freezer or fridge and an external thermostat. At high kräusen, the environmental temperature may need to be lowered an extra degree or two to keep the desired temperature constant. Likewise, the temperature may need to be raised to the target fermentation temperature near the end of fermentation as yeast activity slows.

Because the beer may not exactly match the surrounding temperature, you should have a temperature probe monitoring the beer itself if possible rather than the environment. A stick-on thermometer affixed to the fermenter is a common way to measure beer temperature as it ferments. These are not very precise, but they are inexpensive and give brewers a good idea of the beer temperature to within a degree or two.

BLOW-OFF TUBES

Vigorous fermentations can produce so much kräusen that it rises and pushes out of the fermentation vessel through the airlock. One solution to this is to affix a blow-off tube. A blow-off tube will also remove some of the bitter compounds that get pushed up by the kräusen and cling to the inside of the tube or are expelled into the water lock. If you are brewing a malt-focused beer, this can help you achieve a smoother bitterness. In contrast, if you're brewing a double IPA, you might not want to lose those compounds. If you expect a vigorous fermentation, choose a fermenter with a headspace volume that will let you retain or blow off the kräusen, according to your desires.

CLEAN UP

After any fermentation, but especially lager fermentations, the yeast may need to mop up excess diacetyl. Don't ever rush to separate the beer from the yeast the minute that fermentation is complete. If you are using a diacetyl-prone yeast, don't rack the beer off the yeast until you've sampled it and confirmed that the diacetyl is gone.

STUCK FERMENTATIONS

At some point every homebrewer faces the dilemma of a dreaded stuck fermentation, which is when the yeast ceases activity before all of the fermentable sugars are converted into alcohol. If you are careful and take steps to keep your yeast happy and healthy, however, you can avoid getting "stuck."

A stuck fermentation is often the result of one of three common conditions: improper fermentation temperature conditions, unhealthy yeast (or not enough healthy yeast cells), or a lack of oxygen.

TEMPERATURE

Yeast can be fickle under the wrong temperature conditions, and more specifically they don't like to be too cold or too hot. Yeast suppliers provide temperature guidelines for each of their yeast strains, which are ranges that they have determined in their laboratories as the temperatures that the yeast are able to grow and thrive without going dormant or dying, while producing the best beer. When you're brewing a batch of homebrew, be sure your fermenter is kept in an area that doesn't get too cold, which is a common reason for a stuck fermentation. When brewing beer styles that need to be kept on the cooler side, such as lagers, keep a close eye on the temperature inside your fermenter. If your fermentation starts to slow or stop, you can try warming things up a few degrees by moving your fermenter to a warmer area or with an electric heat wrap around the fermenter to get things moving again.

UNHEALTHY YEAST

One of the most important steps for brewing any beer should always be pitching enough healthy yeast. Without enough healthy cells, the yeast can struggle and even decide to quit. If you are brewing anything with a higher-than-normal gravity, or anything that needs to ferment at a cool temperature, it's a good idea to build up a healthy population of yeast a day ahead of pitching with a yeast starter, or at least pitch more liquid or dried yeast than the recipe might call for. For more information about yeast starters, turn to page 42.

If you have experienced a stuck fermentation, depending on where you are in your fermentation (take measurements with your hydrometer), you can try repitching more yeast. If fermentation stops near the beginning or middle of fermentation, you can pitch another full dose of yeast. If the fermentation stops near the end, try pitching a smaller amount of yeast—about a pint of yeast as a starter. You can also try adding yeast nutrient to be sure the yeast is healthy. Another trick is to kräusen the beer by adding some beer that is in the high kräusen stage of fermentation (36 to 48 hours after pitching for most beers). The rule for kräusening is to add 10 percent of the fermenter volume, or 0.1 part kräusen to 1 part beer.

OXYGEN

In addition to temperature constraints, yeast need oxygen. Aerate your wort well before pitching the yeast, which many beginner brewers do by letting the wort splash when transferring it into the fermenter followed by vigorously shaking their fermenter. A more fail-safe method of aeration, however, is to invest in a simple aeration stone setup. For more information about proper aeration, turn to page 41.

Oxygenating with pure oxygen is ideal, but you can also introduce some oxygen for the yeast by shaking the bucket with the lid on.

TIPS FROM *BYO*

FININGS

"Finings" is brewer speak for flocculants that are used in brewing to clarify suspensions of solids in a liquid, such as trub in wort and yeast in beer. Such solids remain suspended because they are small so they can settle only very slowly, and because they usually carry the same charge and so repel each other and prevent aggregation. Flocculants are generally big molecules, usually carrying some charge and soluble in the liquid medium concerned and are capable of attaching to several suspended particles at the same time, thus making one big particle (floc). Since the settling rate of a particle in a liquid is directly proportional to the square of the particle diameter, a flocculant can rapidly clarify a suspension that otherwise might remain hazy for days or weeks. In the brewery, finings/flocculants are used in two places, namely wort clarification and finished beer clarification, and the types of flocculant used in each of these cases are quite different.

FINING WORT

Kettle finings work only during cooling. In other words, they do not flocculate the hot break (trub), but the material that precipitates as cold break. This improves the colloidal stability of the beer, thus helping to limit haze formation in the finished product. The cold break consists mainly of proteins and protein degradation products. So, the flocculant used for this purpose must be capable of reacting with these particles. Proteins are based on amino acids and are what chemists call "amphoteric," that is to say they possess both basic and acidic properties, and whichever property is dominant will depend upon the pH (acidity) of the medium. In our case the medium is wort, which will be on the acid side (generally less than pH 5.0), and the proteins and protein residue will be mainly positively charged.

The substance that has been found most suited to this is called carrageenan, or Irish moss. It is a linear polysaccharide, linearity generally being a desirable quality in a flocculant molecule. But more importantly it carries a number of sulfate groups that are negatively charged, even at wort pH, and react directly with the positively charged protein molecules.

Carrageenans are obtained from seaweed, notably Chondrus crispus, which is found off the coast of Ireland, and Eucheuma cottonii, a Pacific seaweed. Pure Irish moss is available and works well, or it is also sold in the form of a powder or as tablets. These products come under various names, such as Whirlfloc, Koppakleer, Protofloc Supermoss, and so on. Depending upon the product, you will need to add only a small amount. A 5-gallon (19 L) brew usually requires 2–2.5 g powdered Irish moss, or 0.5–1.5 g refined carrageenan. Most homebrewers don't have the capability to accurately weigh such small amounts, and base additions on the tablet form. Suppliers may also do this, using phrases like "half a tablet per 5 gallons (19 L)." Do not use more than recommended because overdosing (with any flocculant) can result in big fluffy flocs that will retain much of the wort, and if you go even further, high dosages can result in stabilizing the solid suspension, rather than settling the solids.

FINING BEER

In this case we are talking about separating yeast from beer after fermentation. Such fining treatment may be carried out as a stand-alone process or in conjunction with filtration. Since yeast generally carries negative charge at the pH of beer (about 4.0–4.5), a flocculant that is positively charged is desirable. Proteins have a negative charge at this pH, so it is not surprising that the two products most widely used are proteinaceous in character, namely gelatin and isinglass. The basic chemistry of these two compounds is somewhat similar in that they are forms of collagen, but isinglass has a much higher molecular weight than gelatin, which is a hydrolyzed collagen. Therefore, isinglass can form much bigger, faster-settling flocs than can gelatin, which makes it best suited to fining unfiltered draft beers.

Gelatin: Gelatin tends to form much smaller flocs that settle quite slowly so that fined beer in keg or bottle may take days or even weeks to clarify. It can also be used to advantage in aiding filtration, since the small flocs will not blind the filter medium. In that case it is probably best to store the beer for a few days after gelatin addition and before filtration. In either case, best results will be obtained if the fining is carried out with beer that has been racked from the secondary.

You will need 4–8 g (½–1 tsp.) per 5 gallons (19 L). Add it to 100 mL (½ cup) cold water, stir, then heat to just under boiling, stirring continually until the gelatin dissolves. Add the mixture to the racked beer, gently agitate it, and then bottle or keg the beer. If you are priming the beer for bottling, that should be done at the same time as fining. Gelatin is a commodity, and all unflavored gelatins will essentially work the same. Knox is a common brand available in grocery stores.

Isinglass: This is a triple helical molecule and is extracted from the swim bladders of certain fish. It has been used in Britain since the 18th century when it came primarily from Russian sturgeons, but it is now produced from a variety of more common fish. It forms large, fast-settling flocs capable of forming a firm, compact sediment, which makes it an ideal flocculant for cask-conditioned beers. Note that isinglass has a proven foam stabilizing effect on beer.

Isinglass is available from homebrewing suppliers in forms that are either already dissolved or dissolve relatively rapidly, such as Biofine PO19 or Cryofine. Follow the instructions provided by the supplier as products vary. Isinglass is added shortly before kegging or bottling. Note that solutions of isinglass are usually unstable to heat and will degrade if kept at temperatures above 60–65°F (16–18°F), so they are best kept refrigerated unless used immediately.

Whether to fine is a personal choice, and many homebrewers do not find it necessary. After all, in an age when the most popular beer style, IPA, is often so highly hopped that it will always throw a haze when cooled, clear, bright beer is not always as desirable as once was the case. That said, clear beer is still the expectation for most styles.

BOTTLING

After biological fermentation—the conversion of sugars to alcohol and carbon dioxide—has finished, most beers are conditioned for a period of time during which the yeast and other solids drop out of solution and some molecules are taken up by the yeast. After this conditioning time, which may be only a few days for some ales but up to a few months for some lagers, most beers then go to the packaging stage. (In some cases, however, the brewer wants to add additional flavors or aromas post-fermentation. Two examples of this would be dry hopping or adding fruit to make fruit beer.)

Bottling, it must be said, is a fairly time-consuming part of homebrewing. You need to clean and sanitize all the bottles, then fill and cap each one. Bottling is also rewarding, though. When you put those two cases of homebrew in your closet to condition, there's a sense of accomplishment—and anticipation.

RACKING AND TRANSFERRING

Racking is the process of moving beer from one container to another. While many brewers rack certain beers—or even all their beers—to separate them from the particles that settle at the bottom of the carboy, fermenter, or bucket, racking is an essential skill for bottling and kegging.

To really understand racking, think of your beer as something of a shaken-up snow globe. There are all kinds of particles in suspension in the liquid, such as hop material and yeast cells. Over time, those particles settle at the bottom of the vessel. Yeast cells die and fall to the bottom, and other particles settle.

After primary fermentation, you can remove the beer from these deposits as prolonged exposure to the sediments, especially dead yeast cells, can cause off-flavors. This isn't such an important step if you are making a beer that doesn't need extended conditioning. You can simply rack when it's time to bottle or keg. However, a style that needs to condition longer, such as a lager, may need extra weeks of conditioning and should be transferred off of the sediment into a secondary fermenter. There is some controversy regarding how long you should wait to transfer your beer off of the sediment, but it is ultimately up to the brewer.

HOW TO RACK

To rack beer from one vessel to another, you will need a racking cane or auto-siphon and siphon tubing with a clamp that controls the flow in the tube. Siphoning is the process of using gravity to pull a volume of liquid from a higher vessel into a lower vessel—for example, from a carboy on a table to a carboy or keg on the floor. Using an auto-siphon, you will insert the sanitized auto-siphon with sanitized tubing already attached into the higher vessel, ensure the other end of the tubing runs to the bottom of the receiving vessel, and then use a pumping motion to start the transfer. Once the liquid is flowing and the auto-siphon is positioned near the bottom of the top vessel, you do not need to do anything else. The liquid will flow on its own from one vessel to the other.

The process is very similar when using a racking cane and tubing. However, to start the transfer, you will fill your tubing with sanitizer, attach it to the sanitized racking cane (which will sit inside the top vessel in the same manner as the auto-siphon), and then let the sanitizer run out into a small bowl or other waste container before inserting the bottom end of the tubing into your receiving vessel. You can use a clamp on the tubing to stop the flow of liquid if needed. As long as your racking cane is positioned correctly, you will not need to do anything else, and the liquid will transfer.

LEFT: An auto-siphon makes starting a transfer much easier. **TOP RIGHT:** Place your tubing on the bottom of the receiving vessel to minimize splashing and oxygen pickup. **BOTTOM RIGHT:** Try to transfer all the beer but not the sediment layer on the bottom of the vessel.

The most important part of the siphoning step, however, is to be sure that you rack the liquid above the sediment at the bottom of the container. Racking from the bottom of the container will transfer the particles you are trying to separate out from the beer.

Any time you transfer beer from one container to another, you risk oxidation, and oxidation can cause off-flavors. If you are transferring your beer with a basic setup (just a racking cane/siphon setup), prevent as much oxygen exposure as you can by carefully and slowly transferring the beer, being careful not to splash.

If you are transferring to a carboy or another vessel to secondary your beer, choose a vessel that does not leave air at the top of the liquid when you are finished transferring—this is called *headspace*, and leaving that air in the secondary means that you are basically trapping oxygen in an enclosed space with your homebrew. When you add equipment to your homebrewing setup, you can also use CO_2 to prevent oxidation when transferring your beer. If you use CO_2, you can purge the secondary fermenter with the gas before transferring the beer to push the oxygen out of the vessel.

HOW TO BOTTLE

Now that you know how to rack, it might seem like life is as easy as transferring your beer from your fermentation bucket or carboy into bottles. Not so fast! Bottling requires a few additional steps both to ensure sanitation from start to finish and to ensure your beer carbonates. Let's get started.

You Will Need

- 6-gallon (23 L) homebrew bucket (for sanitizing)
- 6-gallon (23 L) homebrew bottling bucket (with spigot)
- Filtered water and priming sugar (see page 53)
- Approximately fifty 12-ounce bottles and bottle caps
- Racking cane, bottling wand, and tubing (often sold together as a kit)
- Bottle capper

1. Get Ready to Bottle

When bottling homebrew, there are two major goals. First, you want to carbonate the beer to the correct level. Second, you don't want to set the beer up to stale quickly by introducing too much oxygen as you bottle. Let's first start with a few words on avoiding oxygen pickup. In short, the longer the beer sits in the bottling bucket, exposed to air, the more oxygen will get into your beer. You don't need to rush to avoid this, but you should have everything ready to go when you open the fermenter and begin siphoning the beer into the bottling bucket.

Assuming your bottles are already clean and free of surface dirt or debris, mix a bucket of sanitizer, such as Star San. Sanitize at least fifty bottles for a 5-gallon (19 L) batch. A bottle-holding device such as a bottling tree can come in handy here, as it's a good idea to let the sanitizer drip out of the bottles as they dry. Then go ahead and sanitize your bottle caps. Last but not least, make sure your bottle capper is nearby. There's not much worse than filling your first bottle and then hunting for the capper.

Make sure your bottles are sanitized before you get down to the business of bottling.

2. Prime Your Beer

To prime your beer, first read the priming section starting on page 53 to figure out how much corn sugar you'll need. Measure the sugar into a small saucepan and add water until the sugar just dissolves. Now boil the priming solution for 5 minutes (you want to sanitize but don't want to darken the sugar extensively). You should then transfer the priming sugar into your sanitized bottling bucket before transferring your beer (it doesn't matter if it's still hot). Then siphon your beer into the same bucket with as little splashing as possible.

Place your bucket of beer on a tall and sturdy surface (such as a kitchen counter) and siphon the beer into the priming sugar in the bottling bucket using a sanitized racking cane or auto-siphon and tubing. The mixture should swirl as the beer flows in, ensuring that it is mixed in thoroughly with the sugar water. If you want to make sure it's all mixed well, you can give the beer a slow stir or two with a sanitized spoon after transferring. (Don't stir too vigorously though as that just allows more oxygen to enter the beer.)

Note: *If you have a kegging system, you can add a squirt of CO_2 to your bottling bucket before you siphon the beer in, and then again once it's full. If you cover the bucket with aluminum foil, you will trap much of the CO_2 and this will provide a partial barrier against oxygen.*

3. Bottle (and Cap) Your Beer

Next you need to move the beer from the bottling bucket to the bottles. Some homebrewers use a racking cane to siphon their beer into bottles. If you have a bottling bucket with a spigot, this stage is easier because you don't need to start another siphon. You'll just attach the tubing to the spigot on the bottling bucket—this is highly recommended! To begin filling, set the bottles on the ground, making sure the surface is light enough so you can see the liquid level as each bottle fills.

If you're using a bottling wand attached to the other end of your tubing, simply place the wand into the first bottle; when the tip pushes against the bottom of the bottle, it will release the beer. When you reach the proper fill, pull the wand away from the bottom of the bottle, and it will stop the flow. If you're using tubing and a clamp to control the flow rather than a wand, the process is similar. Place the tubing inside the bottle at the very bottom, open the clamp to start the flow of beer, and then close the clamp to stop the flow of beer.

The fill level affects how fast the bottle carbonates. Bottles with high-fill levels carbonate slower and may not fully carbonate. Bottles with low-fill levels carbonate faster and may overcarbonate. The "right" fill level is the level most commercial beers are filled to, about an inch below the top of the bottle. When the bottle is filled, remove the tubing or wand. Place a blank cap on top of the bottle. When you're done filling all your bottles, crimp the caps on with the bottle capper, and you're done!

4. Wait, Then Enjoy!

With the right amount of priming sugar in the beer and the beer bottled and capped promptly, the yeast is ready to carbonate your beer. For best results, store the beer relatively warm, but out of any direct light, for about 2 weeks. Around 75 to 80°F (24 to 27°C) would be great for most beers. When 2 weeks is up, take one beer as a test and place it in the fridge overnight. Open it the next day and check the carbonation levels. If it's okay, move the rest of the beer to cooler storage—preferably at refrigerator temperatures, but anything below ale fermentation temperatures will be adequate.

TOP LEFT: You can use the bottom of the bottle wand and/or a clamp to stop the flow of beer to the bottle. **TOP RIGHT:** The space the wand takes up while filling will leave you with an appropriate amount of headspace once the wand is removed. **BOTTOM LEFT:** Place your bottles on a sturdy, level surface when capping. **BOTTOM LEFT:** It will be easy to tell if your bottle is perfectly capped or not–there's no gray area when capping!

PRIMING

For bottle-conditioned homebrew, beer is primed with a small amount of fermentable sugar and then sealed. The carbon dioxide created during the fermentation of this sugar is trapped in the sealed bottle to create carbonated beer. Many older homebrew sources cite a single amount of corn sugar for priming a 5-gallon (19 L) batch, usually ¾ cup. This generally does produce a level of carbonation suitable for English ales. However, for more control over your level of carbonation, you need to consider the residual level of carbonation and then add sugar to reach your target level, which will vary for different beers.

What is residual carbonation? After fermentation is complete, most of the carbon dioxide generated by fermentation will have bubbled out of solution, but a small amount will remain dissolved in the beer. How much depends on temperature. A beer that fermented at a steady 72°F (22°C) is going to contain slightly less CO_2 than one fermented at a steady 68°F (20°C), about 0.80 volumes of CO_2 and 0.85 volumes of CO_2, respectively, because gases are more soluble in colder liquids. However, even at lager temperatures the amount of carbon dioxide retained is below the level of carbonation of English ales.

From the chart below, you can find the residual level of carbonation in your beer, based on your fermentation temperature. Then subtract this amount from your target level of carbonation. Now you have the amount of carbonation you need to generate during bottle conditioning. You can get that amount from the chart (or see the more extensive charts at https://byo.com/resource/carbonation-priming-chart/.

Note: *Dextrose (corn sugar) isn't the only sugar you can use to prime your beer. However it's the easiest to work with and doesn't impart any additional flavors to your beer. For the vast majority of homebrews, dextrose works just fine and there is no reason to use more exotic sugars such as maple syrup, molasses, etc. If you are interested in priming with other simple sugars, such as sucrose, the priming charts at the link above will provide some common alternatives and quantities.*

RESIDUAL CARBONATION AND PRIMING

Fermentation Temperature °F (°C)	Volumes CO_2 (residual)	Priming Sugar (corn sugar) [weight/5 gal (19 L)]	Volumes CO_2 (bottle conditioning)
50°F (10°C)	1.15	1 oz. (28 g)	0.34
53°F (12°C)	1.09	2 oz. (57 g)	0.68
56°F (13°C)	1.04	3 oz. (85 g)	1.02
59°F (15°C)	0.988	4 oz. (113 g)	1.36
62°F (17°C)	0.940	5 oz. (142 g)	1.70
65°F (18°C)	0.894	6 oz. (170 g)	2.04
68°F (20°C)	0.850	7 oz. (198 g)	2.37
71°F (22°C)	0.807	8 oz. (227 g)	2.71
		9 oz. (255 g)	3.05

BOTTLES PER 5-GALLON (19 L) BATCH

Package Size English (metric)	Batch Size, 5 gallons (19 L)	10 gallons (38 L)	15 gallons (57 L)
12 oz. (355 mL)	53	106	160
16 oz. (473 mL	40	80	120
22 oz. (650 mL)	29	58	87
1 qt. (946 mL)	20	40	60
2 qt. (1,900 mL)	10	20	30
5 gallon (19 L)	1	2	3

The table lists how many bottles or kegs of various sizes it would take to package three common batch sizes for homebrewers. Numbers are rounded to the nearest bottle.

KEGGING

The most popular alternative to bottle conditioning homebrew is kegging it in Cornelius (a.k.a. Corny) kegs. These kegs, with the 5-gallon (19 L) size being the most common, are perfect for most homebrew batches, and once you have the equipment, kegging is more convenient than bottling. Although moderately expensive to get started, it takes much less effort to rack a 5-gallon (19 L) batch of beer into a single Corny keg than it does to put it into fifty-plus 12-ounce (355 mL) bottles. Plus, since you force carbonate beer in a keg, you'll have beer that's ready to drink sooner, and you'll be able to adjust the carbonation level as needed if it's too high or too low initially.

HOW TO KEG

Getting started with kegging requires about a $200 investment if you purchase all-new equipment. You can source each of the pieces individually or buy a basic system as a kit from your favorite homebrew retailer.

You Will Need

- 1 or more Cornelius (a.k.a. Corny) keg with rubber O-ring seal (5 gallon/19 L are the most common size, though smaller sizes are available)
- CO_2 tank (10 pounds/4.5 kg is a good size for most homebrewers)
- Dual-gauge CO_2 regulator
- 1 gas-in quick disconnect
- 1 beverage-out quick disconnect
  Note: *Quick disconnects are made for two types of kegs: ball lock and pin lock. The difference is the type of connection posts at the top of the keg, which are used to connect the gas and the beer line to the keg. There are two different types of kegs because originally these smaller kegs were designed to hold soda syrup. The soda companies developed the two types of incompatible kegs to prevent customers from switching between brands.*
- Gas tubing (3 feet/1 m minimum)
- Liquid tubing (5 feet/1.5 m minimum)

- Hose clamps
- A plastic picnic (a.k.a. "cobra") tap, for dispensing
- Wrenches (for ball-lock fittings, you'll need a $^7/_8$-inch deep socket and socket wrench; for pin-lock fittings, a $^{13}/_{16}$-inch pin socket)
- Keg lube

1. Get Ready to Keg

Once you have gathered your essential equipment, it's time to move on to the business of kegging. Make sure your keg and its parts are clean and sanitized, as well as all of your tubing and your racking cane. It's handy to have a spray bottle of sanitizer on hand during kegging to quickly sanitize any small parts (such as an O-ring). Before transferring to the keg, you can also cold-crash your beer if you would like, which will help drop some of the yeast out of suspension. Now, time to transfer.

2. Transfer Your Beer to the Keg

As with bottling, you want to minimize the amount of oxygen the beer is exposed to. When kegging, there is one way that works well. Fill your clean, sanitized keg completely with water, then use CO_2 pressure to completely empty the keg. You now have a keg that contains only CO_2. Rack the beer quietly to this keg, keeping the lid of the keg loosely set over the opening (or cover the opening with aluminum foil). This way you will rack the beer into the keg under a blanket of CO_2. Seal the keg as soon as the beer has transferred and apply CO_2 pressure (only a few psi is needed), and pull the pressure relief valve a few times. Each time, some headspace gas (which will have a small amount of oxygen in it) will exit the keg and be replaced with pure CO_2. A few spurts of CO_2 will adequately purge the headspace and your beer will be less prone to stale quickly.

3. Carbonate

While you can carbonate a keg by priming your beer as you would for bottling, most homebrewers instead choose to force carbonate with CO_2. Carbonating your beer properly is simplest if you think of it as a function of temperature and time. If you set your beer to a particular pressure level at a particular temperature, you can figure out how long it will take for your beer to reach its equilibrium. Most popular brewing software and many websites have this information, including BeerSmith, The Brewer's Friend, and Beer Recipator.

Kegging requires an investment in new equipment, but almost all homebrewers say the time saved bottling is worth it.

You can also try to speed things along by agitating your keg, which is done by setting your regulator to a high pressure and shaking or rolling your keg vigorously. But there are some caveats to this method. While shaking the keg (often called *crank and shake*) is a frequently suggested method for carbonating a keg, it is also not a very exact method. It isn't "wrong" to carbonate this way, and it is fast, but it won't provide consistent results as you would if you hook up your CO_2 using a regulator and exercise some patience. In a homebrew-sized keg, carbonation takes just 3 to 5 days to complete if you simply hook the gas up and leave your beer alone.

4. Store and Pour It Right

Once beer is bottled or kegged, store it so it lasts as long as possible. Your best option is to store the beer in a refrigerator until the keg runs out. Your next best bet—if, say, you want to take a keg off tap to make room for another and you're limited on space—is to store the beer as cool as possible at a steady temperature. The amount of time homebrew will keep is dependent on many factors. These include the level of contamination in your beer, how well you managed to keep oxygen away from the beer, the strength of the beer, and other variables.

Transferring to a keg is as easy as transferring to a bucket or carboy.

As far as managing your draft system's pours, many homebrewers eventually dial in what works for their particular system, keeping in mind variables such as line length and preferred serving temperature. To get started, though, continue reading below.

BALANCING YOUR DRAFT SYSTEM

One of the real pleasures of homebrewing is serving your own beer from your own tap. A properly set-up and maintained home dispensing system allows you to pour correctly carbonated beer that has the appropriate head and appearance for style. However, it can also be the source of frustration if things are not done right. You can end up with a glass full of foam or flat and lifeless beer, depending. Both of these pitfalls can be avoided with a little knowledge and planning.

THE SCIENCE BEHIND THE BUBBLES

During fermentation, one molecule of glucose is broken down into two molecules of ethanol and two molecules of carbon dioxide (CO_2). The CO_2 that is produced is soluble in beer and results in residual carbonation.

Shaking or rocking a keg can help speed up carbonation, but we recommend setting the psi and letting time do the work whenever possible.

All beer contains at least some dissolved CO_2, and most styles are additionally carbonated, whether by fermentation of added sugar or by "force carbonation" with additional CO_2.

The total amount of CO_2 dissolved in the beer is measured in volumes, which is the volume the gas would occupy if it were removed from the beer and kept at standard temperature and pressure—STP—32°F (0°C) and 1 atmosphere of pressure—divided by the volume of the beer. This is also used to describe the carbonation level of a beer. For example, American lagers generally are carbonated to about 2.6 volumes of CO_2. Less carbonated styles, such as many British ales, can have a carbonation level as low as 1.2 to 1.3 volumes, while some sprightly German wheat beers may be carbonated to levels above 4.0 volumes.

The solubility of CO_2 increases as the temperature decreases. There is also some decrease in solubility as the specific gravity increases, but the effect is small and can be disregarded for the gravity of beer. The solubility of CO_2 also increases with increasing gas pressure. In order to achieve the correct volumes of gas in beer, it must be stored under pressure (in either bottles or kegs).

MAY THE FORCE (CARBONATION) BE WITH YOU

The correct procedure for force carbonating beer to the appropriate carbonation level is outlined in the previous section. To summarize, you can set the regulator pressure to the appropriate level (from the formula below, the carbonation chart on page 53, or brewing software) and let the beer carbonate over a period of several days. Or you can use the crank-and-shake method (which is less exact but requires less time) of setting the regulator to a high pressure and shaking the keg vigorously for several minutes and repeating several times over a period of a couple of hours.

The formula for setting the regulator to the correct pressure—P (in pounds per square inch, or psi) for the desired level of carbonation, V (in volumes of CO_2) at a beer temperature of T (in °F)—is $P = -16.6999 - (0.0101059 \times T) + (0.00116512 \times T2) + (0.173354 \times T \times V) + (4.24267 \times V) - (0.0684226 \times V2)$.

PROBLEMS DOWN THE LINE

Assuming the beer is carbonated to the appropriate level, it still has to make its way from the keg, through the line, out the tap, and into the glass—and this is where problems can occur.

If the dispensing pressure is too low, the beer will pour too slowly and excessive foaming can result, to the point where little beer and mostly foam ends up in the glass. Furthermore, over time the beer in the keg will lose carbonation as more CO_2 comes out of solution as it attempts to achieve equilibrium with the headspace. At extreme underpressure, the beer can become nearly flat. If the dispensing pressure is too high, it, too, can result in excessive foam from the beer pouring too quickly from the tap. With time, the beer will become overcarbonated as more CO_2 goes into solution, further complicating the situation.

At lower than the correct pressure, in addition to low carbonation, the line will tend to collect bubbles and pockets of CO_2 where it has come out of solution, especially just above the keg and behind the faucet, as well as in places where the temperature is warmer. These pockets will become larger the longer the time

Foamy pours are a sure sign that your draft system is out of balance.

period between dispensing beers. The first beer will have a shot of foam, followed by clear beer, followed by more foam. After pouring a few beers, the problem may dissipate, only to return again after a rest. Low-pressure problems also tend to show themselves early when a keg is nearly full.

At higher than optimum pressure, there will be overcarbonation and symptoms similar to those that occur at low pressure. The difference is that they tend to appear and grow worse as the keg is emptied. If a fresh keg is foamy, the odds are that it is not an overpressure problem. The reason is that, as the keg empties, more CO_2 occupies the larger headspace as the difference between the equilibrium pressure and dispensing pressure increases. Again this causes excessive foaming when the beer is first dispensed, until there is less CO_2 and more beer in the line.

A MATTER OF BALANCE

Calculating the correct dispensing pressure and making changes to the system is known as *balancing* and is critical to pouring a perfect beer. Balance is not only dependent on the carbonation level and the temperature of the beer, but several other factors also enter into the equation. These include the overall height difference between the keg and the tap, the length and diameter of the dispensing line, and the type of tap being used. Changes to any one of these will change the balance of the system.

Between the keg and the tap, there is resistance to the flow of the beer. Gravity (the difference in height) accounts for 0.5 psi per foot (11.3 kilopascals per meter), a positive value if the tap is located above the keg, negative if the tap is below it. A standard beer faucet has a resistance of 2 psi (13.8 kPa); the shank adds another 1 psi (6.9 kPa). A picnic or "cobra" tap has a resistance of about 0.5 psi (3.4 kPa). Additionally, the beer line itself offers the following resistance based on the inside diameter. These figures are for flexible vinyl beverage tubing:

$3/16$ in. (4.75 mm) inside diameter (ID): 3.0 psi/ft. (67.9 kPa/m)
$1/4$ in. (6.35 mm) ID: 0.8 psi/ft. (18.1 kPa/m)
$5/16$ in. (7.94 mm) ID: 0.4 psi/ft. (9.0 kPa/m)
$3/8$ in. (9.53 mm) ID: 0.2 psi/ft. (4.5 kPa/m)

The material out of which the beer line is made affects these ratings. As such, your beer line may vary from these numbers. If your tubing is meant specifically for beer lines, its resistance might be found on the manufacturer's page online. If not, these numbers are a good starting point for your calculations, which may have to be adjusted by trial and error later.

Finally, some additional pressure is necessary to achieve a proper flow rate. The generally accepted desirable pour rate for beer is considered to be 1 US gallon (3.8 L) per minute or 1 US pint (473 mL) per 7 to 8 seconds. For most systems, a value of 5 psi (34.5 kPa) is sufficient for balancing calculations.

Assuming that the other values remain the same, the easiest way to balance the system is to adjust the line length so that the total resistance of the system equals the carbonation pressure minus the required 5 psi (34.5 kPa) for a proper flow rate. Round the result to the next highest foot (0.3 m).

For example, for a pale ale that is carbonated to 2.3 volumes of CO_2 at 46°F (8°C), the correct carbonation pressure (from the force carbonation formula) is 13 psi (89.6 kPa). The beer is dispensed through a standard shank and beer faucet at a height of 2 feet (60.9 cm) above the center of the keg.

Here are the calculations for the required length of $3/16$-inch (4.75mm) diameter beer line in order to balance the system:

Gravity resistance: +2 ft. (60.9 cm) × 0.5 psi/ft. (11.3 kPa/m) = 1 psi (6.9 kPa)
Shank resistance: 1 psi (6.9 kPa)
Faucet resistance: 2 psi (13.8 kPa)
Fixed resistance of the system (not including the line):
        2 + 1 + 1 = 4 psi (13.8 + 6.9 + 6.9 = 27.6 kPa)
Carbonation pressure of the beer (2.3 volumes of CO_2 at 46°F/8°C): 13 psi (89.6 kPa)
Pressure required to dispense beer at 1 gallon (3.78 L)/minute: 5 psi (34.5 kPa)
Pressure needing to be balanced: 13 − 5 = 8 psi (89.6 − 34.5 = 55.1 kPa)
Resistance to be supplied by the line: 8 − 4 = 4 psi (55.2 − 27.6 = 27.6 kPa)
Resistance of $3/16$-inch (4.75mm) ID beer line: 3 psi/ft. (67.9 kPa/m)
Length of $3/16$-inch (4.75mm) ID line required to achieve 8 psi (55.1 kPa) resistance:
        $4/3$ = 1.33 feet (40.5 cm)
Rounded to next highest foot (0.3 m): 2 feet (61 cm)

The perfect pour is always attainable if you pay attention to a few key variables.

Therefore, 2 feet (61 cm) of ³⁄₁₆-inch (4.75mm) ID diameter tubing will balance this system for the example beer. **Note:** *This length seems short by homebrew standards because 5 psi is a higher overpressure than most homebrewers use. Lowering the dispensing pressure to 0.5 to 1 psi will result in a line length more in line with usual homebrew setups. Experiment with flow rates to find one you like.*

ACHIEVING NEW BALANCE

Of course you may choose to serve a variety of styles at different carbonation levels and perhaps at different temperatures. This will affect the system balancing equation somewhat. You can recalculate the new carbonation pressure and line length necessary to balance the system, and adjust the dispensing pressure and replace the line with the proper length. If the difference is small, you may choose to ignore the slight imbalance; balancing a system does not require extreme precision. Or you may use what is called a *choker*, a short length of smaller diameter line installed at the faucet shank. A more elegant solution is to purchase and install a line restrictor, which allows you to vary the flow of beer through the line. These devices are available from draft beer equipment suppliers.

In reality, you will likely balance your kegging setup once, then use the same beer line length for all your beers, provided they are all carbonated to similar levels. If you change your setup—for example, because you bought or built a new kegerator—you will need to balance it again.

Keep in mind the number yielded from the calculations above is only as good as the quality of the estimates of resistance for your equipment. If your beer line has more or less resistance than the figures quoted here, your calculations will be off. However, unless the particulars of your system are far different from average, the equations above should get you within the ballpark. And, when you do the calculations, save them in your brewing notebook for use later if you change your dispensing setup and are using the same type of beer line.

When setting up your dispensing system, keep in mind that beer line is cheap. Do the calculations for balancing the system, but then start with a length of beer line that is substantially longer than the equations suggest. Try this and shorten the line length, if needed. On a system with more than one beer line, you will only need to do this once. When you find a beer line length that works, simply cut all your lines to that length. (This assumes all your beers are being conditioned at the same temperature and pressure.)

Try to maintain an even temperature throughout the system. Carbon dioxide tends to come out of solution and collect in warm places, especially near the tap. This is why you may want to discard the first small amount of beer and foam that is in the line after the system has not been used for a while. If the taps are enclosed in a tower, insulate the lines or provide them with a supply of cold air.

Beer that has been recently carbonated by cranking up the regulator pressure and shaking the keg may pour with a lot of foam because the carbon dioxide has not dissolved evenly into the beer. Whenever you are balancing your system, use a keg of homebrew that is fully conditioned. Otherwise, you will get a bad pour even when the system is balanced. Beer line deposits also can increase restriction and cause dispensing problems. This is another reason to regularly clean the lines and taps. And finally, keep your glassware clean, well rinsed, and free of soap deposits—and store them at room temperature. Frozen glasses or mugs will cause foaming and greatly reduce beer aroma and flavor. The proverbial frosty mug is a gimmick that does not improve the quality of the beer.

BREWING INGREDIENTS

MALT

Choosing the proper malts for your homebrews can seem daunting. There are so many options! Let's run through some of the most important aspects of this complex ingredient. We'll focus on how malt is made, the various types of malt, how to taste and select malt, as well as storing and using malt—all keys to perfecting your brewing results.

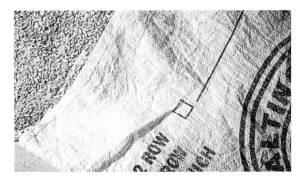

Today's malts are generally well modified and only require a single-infusion mash.

THE PROCESS OF MALTING IN A NUTSHELL

For the purposes of brewing, the word *malt* can refer to any raw grain that has been germinated and dried for enzyme development. These enzymes are necessary to convert the starches in grain into sugar. The process also shelf stabilizes the product, which is crucial for long transport and storage. There are typically three or four steps depending on the type of malt made: steeping, germination, and kilning (or drying). The fourth step, roasting, is used to make specialty malts such as caramel and other darker malts.

STEEPING

The steeping process typically lasts around 48 hours. During this time, the grain is soaked and drained in intervals—often 4 to 8 hours at a time—depending on the variety of barley and how well it takes up water. As the grain absorbs the water, the total amount of moisture in the kernel jumps from around 12 percent up to 44 percent. With this additional water, the critical ingredient for life, the barley seed, thinks, "I should grow into barley." The current available enzymes inside stimulate the embryo, which wakes up and goes to work. It creates more enzymes, which in turn break down protein around the starch inside. Hormones inside the kernel then recognize that they have water and food available and begin to activate growth of the acrospire, the main plant shoot inside the husk, and sprout rootlets. This little white rootlet sprout just visible out of the barley is known as chit, and the barley is referred to as chitted at this point.

Grain is steeped so germination can begin.

GERMINATION

As soon as the barley is chitted and at the correct moisture uptake level of around 44 percent, it is moved from the steeping tanks to long beds referred to as germination tanks. Here the barley sits for 4 or 5 days and the maltster watches the chit grow into rootlets. The grain is turned during germination to keep the barley from growing together and to maintain consistent growth throughout the bed. If the grain was left alone during this time, it would create one giant mass of barley. This giant mass, sometimes called *felting* as it resembles the fibrous material of the same name, is the maltster's worst nightmare. If this occurs, the barley must be manually separated prior to moving to the kiln.

Germination in progress–before the malster halts it by kilning.

During germination, modification occurs. Most malt on the market is fully modified, referring to the germination process completely breaking down the grain's proteins and carbohydrates until the starch reserves are easily accessible by enzymes, allowing us to turn them into sugar during the mashing process when brewing.

KILNING (AND ROASTING)

If the maltster were to let germination continue, malthouses would be greenhouses growing barley and not making malt! Before the barley plant can use up all of the good stuff inside the seed that we want, such as the starch and enzymes, they need to stop germination. Germination is halted by kilning, a drying process. The malt is exposed to heat for anywhere from a few hours to a few days to lower the moisture in the malt, typically to around 4.5 percent. Time and temperature will be the maltster's main tools for determining the final malt product they make.

Whereas kilning typically uses temperatures between 160 and 220°F (71 to 104°C), roasting often uses temperatures from 160°F (71°C) all the way up to 750°F (around 400°C). This temperature range allows for darker malts and a larger variety of color and flavor. Roasting uses a number of different techniques and also includes products made from raw grain that are not germinated and kilned into malt. Caramel malts, which are typically steeped to convert the starch into sugar inside the kernel, are also roasted. After steeping they are roasted at high temperatures, which creates a hard glassy bead of sugar inside the grain.

Through the malting process, the malster helps the barley create the majority of enzymes used in mashing and brewing. Keep in mind this can be attained with almost any variety of grain, not just barley. Wheat, oats, and rye are also popular malt options—and in some beer styles, the majority of the grain bill. Besides malted grain, brewers also use a smaller amount of different sugars and adjuncts. Though this is more fully discussed on page 72, it's worth mentioning here as broadly speaking many adjuncts for brewing are unmalted, or raw grain. This includes unmalted varieties of barley and wheat, as well as rice, oats, rye, corn (maize), or any other type of grain that has not

Malt is heated in large, entire-room kilns to stop the malting process. Some malts are then heated further to create crystal and roasted malts.

been malted. These grains may have been processed into flakes, torrified (or popped), ground into flour, or you may use them in their raw kernel form depending on the recipe. No matter which form you use, these grains typically have less enzymatic power, which can lead to less sugars and flavor contribution without high enzymatic grain present, or enzymes added.

BASE MALTS

The majority of malts that are made in a kiln are base malts. These malts provide the majority of brewing starches and the enzymes to convert those starches into sugar. The two most common base malts are typically referred to as two-row (2-row) and six-row (6-row). These names are so generic that they often cause a lot of confusion for homebrewers. Two-row and 6-row are the general classifications given to many different varieties of barley, based on their growing pattern. Six-row barley grows in three rows on alternating orientations up the stalk, whereas 2-row barley only has two opposing rows up the stalk. The industry adopted any malt that was light in color, high in enzyme activity, plump, and starchy as 2-row or 6-row base malt. Some companies provide a further descriptor.

Beyond standard 2-row and 6-row, base malt becomes more interesting. Often certain

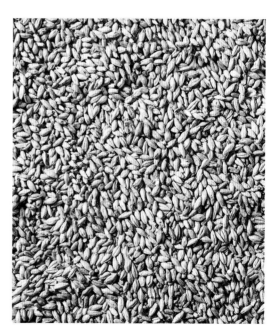

Base malts form the foundation of every beer recipe.

base malts are considered specialty or traditional base malt styles. A specialty base malt would be something such as Gambrinus's ESB malt. It's based on traditional English pale ale malting processes, but they've adapted it to their barley varieties and modern malthouse. Here's a look at common names for a variety of base malts. All of the malts in the Traditional Base Malts chart below can be used to make up 100 percent of your grain bill.

TRADITIONAL BASE MALTS

Malt	Description	Other Names
2-Row Malt	The most common malt in the industry. Rarely darker than a pale yellow 3°Lovibond in color and high in enzyme activity. The flavor is typically sweet and neutral.	2-Row, Base, Brewers, Domestic, 2-Row Pale
6-Row Malt	Grown in the United States almost exclusively. High enzyme activity makes it the preferred base malt with adjuncts. Sometimes has a grainy flavor or more protein than other bases, but nearly identical to 2-row.	6-Row, American, Base, Domestic
Pilsen Malt	The traditional choice for light-colored lagers. It is typically the lightest base malt available and has a slight grassy flavor.	Pilsner, Pilsener, Pils, Lager
Pale Ale Malt	The "grown-up" version of 2-row. Its darker color provides a much bigger and more complex flavor. Extremely popular for English ales and full-bodied American beers.	Pale, ESB, Special Malt, Best Malt, Extra Pale
Vienna Malt	A base malt with a slight toast and nutty flavor. It creates a clean finish and is excellent in lagers when some complexity is needed.	Vienne
Mild Malt	Often used in conjunction with other base malts due to higher color and lower enzyme activity. Adds a malty sweetness.	Mild Ale, English Ale
Munich Malt	Comes in many colors and varieties. Only varieties 10°L or lower in color should be used as a base malt. Anything darker has little to no enzymes. Adds maltiness and is popular in amber lagers.	Munich 5°L, American Munich, Munich 10°L
Wheat, Oat, Rye, and Other Malts	Any grain can be made into a base malt, but barley is by far the most popular.	N/A

SPECIALTY MALTS

These malts are often the fun part of a recipe as they bring the flavor. If base malt is the canvas, then specialty malt is the paint. Anything that won't make up the majority (more than 50 percent) of your beer recipe can be referred to as a specialty malt. Many of these malts spend a longer time in the kiln, at higher temperatures, or get roasted to add depth, complexity, and flavor to the resulting beer. Many of these malts often get unique and trademarked names, as each malthouse has different techniques, tricks, and secrets it employs in making them. The Specialty Malts chart below shows some of the most common types of specialty malts.

Did you find the flavor in your barley you were looking for? Looking for something different? Maybe it's time for malted quinoa. Caramel sorghum? Black roasted millet? Unlike new, dynamic, and interesting hop flavors that the industry begs for, no one seems to be creating

Specialty malts have a wide range of flavors. Experimentation is key to finding what you like!

the same frenzy over malt. With a little more understanding of what can be done, and what flavors can be created, the world of malt can create a variety of new flavors with new grains using new processes. As homebrewers, we love to play with flavor. It's up to us to gain a good understanding of what the malt we're using tastes like and what attributes we want to see in the future.

SPECIALTY MALTS

Malt	Description	Other Names
Acidulated Malt	Traditionally made by spraying sour wort on base malt to encourage lactic acid bacteria growth. This adds a sharp, tart flavor that can accentuate the underlying maltiness.	Sauer
Amber Malt	Malt roasted at a low temperature to provide a sharp bready flavor. Used traditionally in English ales.	Toasted, aromatic
Biscuit Malt	Like amber malt, but roasted at higher temperatures. The color is typically a golden orange 20 to 40°L, like amber, with more distinct biscuit, nutty, and bread crust flavors.	Dark amber, Munich 30°L Briess Victory
Black Malt	Malt roasted at a very high (combustion) temperature, which leads to dry, burnt, and bitter flavors. The malt is at risk of catching on fire and is quenched with water during the roasting process.	Black, black patent

Malt	Description	Other Names
Brown Malt	Often roasted at low temperatures for an extended period of time until color reaches around a milk chocolate brown of 100°L. Often has coffee and nutty flavors.	Coffee, porter
Caramel Malt	Caramel refers to the dominant flavor of these malts. Often called crystal in reference to the crystallized sugar inside. Sweet and in a wide range of colors and flavors.	Crystal, cara
Chocolate Malt	Comes in a variety of colors and flavors. Malt is roasted and the flavors created most often are similar to coffee and dark chocolate.	Pale chocolate, dark chocolate, Weyermann Carafa
Dextrin Malt	A low-color kilned malt, similar to a base malt, but with no enzymatic power and glassy inside. Dextrin gets its name from the type of sugar inside. It contributes a very pale yellow color to beer and is often lighter than 2-row and 6-row base malt. Enhances foam and body.	Carapils, Briess Carapils
Hybrid Malt	No malt is called hybrid. This just describes the dual use of caramelizing in a roaster and then further roasting at higher temperatures. Almost always described by a unique product name.	Briess Carabrown, Special B, extra special, Simpsons Double Roasted Caramel (DRC), etc.
Roasted Barley	Not technically malt, it is an important specialty ingredient. Barley is added "green" (unmalted) to the roaster, which lends itself a milder burnt flavor profile than black malt.	Black roasted barley, roast barley, black barley
Smoked Malt	Typically a base malt that has had smoke flavor added to it (making it a specialty malt). This malt has an intense smokiness and is typically used in restrained quantities.	Rauch
Specialty Rye, Wheat, Oat, and Other Grain Malts	Any malted grain can be used to make the specialty products described above. Barley is the most popular grain used for brewing, but numerous wheat, rye, and oat specialty products exist as well.	Chocolate wheat, caramel rye, Simpsons Golden Naked Oats, Briess Caracrystal Wheat

SELECTING MALT: TASTE IT!

As the recipe designer and brewer, you need to take it upon yourself to do as chefs do: taste your ingredients, take notes, understand what you're brewing with, and know your options. This is perhaps most critical with malt, where subtle nuances can change everything. Just as hops change year to year, so does barley grown for malting. They're both agricultural products, and they can change dramatically, especially as they age.

Flavor should be your biggest concern when choosing malt. Go to your local homebrew shop and taste everything. Smell it, chew it. You can't do that when you buy online, and it can make a big difference in the malt you buy. Another benefit of a local homebrew shop is the service. If you go in and don't like a malt you had in your recipe, look around, try things out, and ask for a recommendation.

Tasting a malt before you buy it and brew with it is critical. If you don't like how it tastes, you probably won't like how it tastes in the beer. A chew test can also indicate age, freshness, and potency, and it will bring out the most distinct flavors of the malt. Performing a taste test is simple:

1. Put a quarter ounce of malt in your hand.
2. Check the aroma. Does the malt smell okay?
3. Chew the malt for at least 1 minute. The flavors that will develop in your beer will slowly come out.
4. Swallow. Does it still taste good in your mouth? The aftertaste can tell you a lot.

One other thing a taste test can tell you is if the malt is slack. Slack malt is malt that has gone soft. This occurs as moisture enters the grain through humidity in the air or wetting. When you chew the malt, it will feel soft and will have lost some of its flavor. Prior to brewing with your beer, this is the worst thing that can happen to your grain. There are a variety of associated issues, including poor milling, mold that can lead to musty flavor and aroma, a turbid muddy wort, and a loss of efficiency in the mash tun. Usually this is a result of poor transport or storage, something to keep in mind when storing your malt.

Note: *Do not refrigerate or freeze your malt; it may cause it to go slack prior to milling due to moisture buildup as it is brought up to temperature.*

Once you've selected and purchased your malt and you are home, you can also do a wort evaluation. Evaluating the wort or liquid-based characteristics of a malt can be simple as well. Numerous scientific methods used in laboratories, breweries, and malthouses can provide the brewer with sensory information on a malt. However, a few simple methods for wort sensory analysis can be used at home. The first is known as the coffee pot method:

1. Finely mill 2 ounces (57 g) of malt in a clean coffee grinder.
2. Add ground malt to a coffee filter and place it in your coffee maker (it's best to use a clean pour-over style brewer for this unless you keep your electric coffee pot very clean).
3. Add 8 ounces (1 cup/0.23 mL) of water to coffee maker to make the wort as you would make a cup of coffee.
4. Let cool and pour in clear cup to see color attributes.

This method can be transferred to many other beverage-making techniques that strain the basis of the beverage. If you own a tea strainer, or French press, you can follow the filter method:

1. Heat 8 ounces (1 cup/0.23 mL) of water to 160°F (71°C).
2. Grind the malt into coarse flour and add to hot water in a French press or tea filter.
3. Let stand for 15 minutes.
4. Remove malt flour and let wort cool prior to sampling.

Tasting the wort will give you an even better understanding of the qualities the malt will add to your beer. Take notes! Look at the color, check the aroma, and taste it. Many flavor wheels can be found online for beer, which can be used for tasting wort as well. Also, check the manufacturer's description of the malt—do you taste what they taste? Use their descriptors and add words such as *slight* or *strong* so you remember what you tasted and smelled.

STORAGE AND FRESHNESS

Malt, when stored properly, has been designed to last a long time. The process reduces moisture, the main enemy of malt. Whether your malt is milled or whole grain, keeping it in an airtight container is key. It is often said that milled (or ground) malt expires fast and should be used immediately. This is true if your storage is less than ideal. Humidity penetrates your grain much easier and faster once it is milled. Moisture is the enemy, and your grain will become slack. However, if you're able to keep the malt cool, as well as stored in an airtight container, your milled grain can last for months. When the temperature gets warm or the temperature fluctuates, though, the malt degrades and may take up moisture quicker. It's critical for grain to be shipped properly or it can become musty, moldy, and slack quite easy. So what are the ideal conditions for storing your malt?

1. Airtight container or original sealed sack
2. Between 50 and 70°F (10 to 21°C) with little fluctuation in temperature
3. An area free of pests such as mice and insects
4. Low to no humidity

Most manufacturers of malt provide a "best by" date for their malt. Using a lot number on your malt, you may be able to determine when your malt was manufactured by contacting either the distributor or manufacturer. Many manufacturers believe that if stored in ideal conditions, malted products can last years beyond their best by date. Let's look at an average of best-by dates for a few brewing product categories in in the Malt Expiration Chart below.

MALT EXPIRATION CHART (AVERAGES BASED ON INDUSTRY DATA)

Malted Product	"Best Used By" Parameters
Whole Grain Base Malt	18 months to 2 years
Milled Base Malt	Ideally use within 6 months, but can last up to 18 months with only aroma loss
Whole Grain Specialty Malt	18 months to 3 years, depending on type
Milled Specialty Malt	3 months to 18 months; some specialty grains are more susceptible to aroma loss and moisture issues when milled

ADJUNCTS

In brewing, adjuncts broadly refer to fermentable ingredients that have not been malted—although not all brewers define adjuncts the same way, and in this section we will also discuss some malted adjunct ingredients such as wheat and rye. It's useful to think about adjuncts divided into two broad groups: kettle adjuncts and mashable adjuncts. Kettle adjuncts, such as honey or candi sugar, contain fermentable sugar and are often added to the kettle in the boil (or sometimes directly to the fermenter). Mashable adjuncts contain starch. Since this starch needs to be converted to sugar before it can be used by brewer's yeast, these adjuncts must be mashed, which means that enzymes degrade the starch to fermentable and unfermentable sugars and dextrins.

Note: *Most adjuncts—including rice, corn, and kettle sugars—contain very little protein, and they are reluctant to yield the protein they do have during mashing. So they also can be considered in terms of their ability to dilute the protein in a wort made from low-protein adjuncts and malted barley. All the protein in this wort comes from the barley, so adding a source of extract that carries no protein effectively dilutes the total protein in the wort. Protein in barley can cause haze. People generally prefer beers to be crystal clear, and they expect that clarity to last for months. So by diluting protein with the proper amount of adjuncts, brewers can increase clarity and stave off the onset of chill haze.*

When brewing with low-protein adjuncts, brewers must take care not to dilute the malt's soluble nitrogen too much or a wort may be produced that lacks enough amino acids. Yeast cells need simple soluble amino acids in order to grow. Nutrient deficiency can result in poor yeast performance and off-flavors. Most of the precursors to stale flavors in beer are derived from malted barley, so diluting the malt with a non-malt adjunct may reduce stale flavors.

Before the enzymes in the mash can break down the starch in the cereal, whether it's corn or malted barley, the starch must be gelatinized. Because starch is a mixture of chemical compounds, it forms a thick gel before it forms a solution. The way the starch is packed into the endosperm affects the temperature at which it will form gel.

Some mashable adjuncts have low gelatinization temperatures, and some have high gelatinization temperatures. This has a tremendous effect on how we use the adjuncts. The starch in unmalted barley will gelatinize at 140 to 143.5°F, with the starch in malt slightly higher at 147 to 152.5°F. The starch in wheat gelatinizes at 125.5 to 147°F, so when it is added to a malt mash, it will gelatinize along with the malt starch. Both corn (at 143.5 to 165°F) and rice (at 142 to 172°F) have high gelatinization temperatures and require a separate heat treatment. Usually corn and rice are mashed separately, along with some malted barley (10 percent), and then boiled in a cereal cooker. They are held for a while as they are heated at a temperature of 158°F to allow malt enzymes to act on the starch and make it less viscous. The cereal mash is then added back to the main malt mash at a controlled rate to raise the temperature of the main mash to its various enzyme rests.

KETTLE ADJUNCTS

Many adjuncts already contain soluble sugar and do not need to be mashed. These adjuncts are added to the wort during the boil and are called kettle adjuncts. This group includes a wide variety of sugars and syrups. Syrups may be produced directly from sugar beet or cane, or extracted from corn or wheat starch. They may be pure glucose (dextrose) or a mixture of glucose and fructose (invert sugar). Or they may contain maltose, maltotriose, and large dextrins. Kettle adjuncts are used in small amounts, typically less than 10 percent of the grain bill. But like cereal adjuncts, they can be used in much higher amounts.

Most types of honey have a delicate flavor, so adding small amounts—less than 5 percent of the fermentables—will have little effect. If you want a big honey flavor in your homebrew, you can add up to 30 percent honey. If you use more than 15 percent, however, add some yeast nutrients. If you are brewing with molasses or maple sugar for the first time, go easy. Both are available in a variety of forms that are more or less flavorful.

Other adjuncts mainly just boost the strength of the beer. Cane sugar, corn sugar, Belgian candi sugar, corn syrup, and rice syrup add little flavor when used in small quantities (around 10 percent).

They boost the amount of fermentable sugar in the wort without adding protein and color to the beer. Over about 10 percent, these sugars can lend a solvent-like, highly alcoholic taste. If you use more than 30 percent in a recipe, yeast nutrition can suffer, especially in beers made from malt extract. Let's take a closer look at some of the most common brewing sugars.

Corn sugar or table sugar: Corn sugar is the more common brewing sugar, which can be used for priming (see page 53) as well as a component to increase ABV and lighten the body of a beer. Table sugar has a similar effect, though its contribution to gravity is higher (1.046 compared with 1.037 for 1 pound dissolved in 1 gallon of water).

Invert sugar: A mixture of glucose and fructose that is produced by hydrolysis of sucrose. Invert sugar is more soluable than cane sugar, and different grades of invert sugar can lend flavors and colors to a beer. It is often seen as a syrup (such as Lyle's Golden Syrup) but can be added as blocks.

Rice syrup: This can be used similarly to corn sugar—to prime as well as boost ABV/lighten body.

Molasses or treacle: A strong-flavored by-product of refining sugarcane or sugar beets into sugar, molasses ranges in color from light to dark. It is typically added in small amounts to the boil. Steer clear of varieties that have been sulfered.

Syrups such as honey are generally added to either the kettle or the fermenter. There's no need to mash them.

Belgian candi sugar: Beet sugar that has been caramelized to produce a dark color and rich flavor, Belgian candi sugar is most recognizably used in its namesake Belgian-style beers. It can be found in varying colors and is most often added in the second half of the boil to prevent further caramelization.

Lactose: Lactose is an unfermentable sugar that is added to boost smooth, sweet flavor in a beer—most frequently milk stouts.

Honey: A high-fructose mixture of sugars that is highly fermentable, honey can come in a wide variety of flavors depending on where bees source their nectar. This character can carry over into a finished beer. Honey can be added to the boil or to the fermenter in varying amounts depending on the character you desire in the finished beer. If you choose to add it to the fermenter, look for a variety that has been pasteurized, and add it either in the secondary or at the very end of primary fermentation to retain its aromatic qualities.

Maple sugar: Made up of mostly sucrose, maple syrup or maple sugar will readily ferment in primary fermentation though it will lose a lot of its characteristic maple flavor. It is better to add maple in the secondary or at the tail end of primary fermentation. If you are buying syrup, make sure you are buying pure maple syrup with no additives.

Maltodextrin: A mostly unfermentable sugar that is added to boost mouthfeel and body in beer, maltodextrine contains a small amount of maltose.

For a complete listing of color and gravity for a wider range of adjuncts, visit: www.byo.com/resource/grains.

MASHABLE ADJUNCTS

Mashable adjuncts can be further divided into two groups, depending on whether the adjunct has the enzymes it needs to break down starch. Malted adjuncts, such as malted wheat or malted rye, contain enzymes; other adjuncts, such as corn or rice, lack them. They rely on the fact that malted barley has a surplus of enzymes, enough to convert the starch of both barley and adjunct.

The degree to which we can use unmalted adjuncts without experiencing difficulties depends on the base malt and the mashing regime. With a multiple-temperature mash, American 6-row malts can tolerate up to 50 percent adjunct, and American 2-row can tolerate up to 30 percent. British malt used in a single-infusion mash can tolerate up to 20 percent.

Flaked and torrified grains are not malted and do not contain the necessary enzymes to convert starch. Flaked grains are made by treating the cereal with steam and then crushing the grain between hot rollers. Common brewing grains in flaked form are oats, rye, corn, and rice. Torrified grains are made by heating grains to a temperature of 500°F (260°C) until they pop, like puffed wheat. Let's take a look at a few of the most common adjuncts you'll see in a mash.

Rice: Rice has a high gelatinization temperature and must be boiled prior to use. Some brewers boil the rice under pressure to increase the temperature. Rice has the highest starch content of all the cereal adjuncts and may yield as much as 90 percent extract efficiency.

Corn: Corn is used by brewers in two main forms: milled grits or flakes. Corn grits are the most widely used adjunct by commercial brewers in the United States and are an important adjunct in Great Britain, where it is called maize. Grits are produced from yellow and white corn (mostly yellow), which is milled to remove the bran and germ. Grits are widely available and require a cereal cooker and separate boiling step similar to brewing with rice. Corn flakes resemble the breakfast cereal and can be used directly in the mash. They can be milled with the malt or crushed by hand and mixed with the grain. Though a box of Kellogg's contains additives that are best left out of homebrew, you can find bulk corn flakes at health-food stores or homebrew shops that are additive-free and sometimes sweetened with malt extract.

Corn and rice are both used in the production of American pilsners. In some of these lagers, the cereal may represent up to 50 percent of the total extract. If you are trying to brew a light-colored beer, using corn or rice will allow you to brew a beer that is lighter in flavor and color than an all-barley malt version of that beer. Typically, brewers use up to 30 percent of these adjuncts.

Unmalted barley: Unmalted barley is used as an adjunct in several major breweries around the world. It is significantly cheaper than malted barley and can be blended at up to 50 percent, provided the enzyme levels are high and an extensive temperature-profile mashing schedule is used. It is difficult to mill as the kernels are extremely hard. It contributes a large amount of beta-glucan to the wort (beta-glucan, a compound normally reduced during malting, can improve foam stability, but in excess it impedes lautering).

Sorghum: Sorghum, also called millet, is the fifth most popular cereal crop in the world. It is used as the base grain in several native African fermented beverages and is used by brewers in Africa and Mexico as part of the grist in their lager-style beers. Sorghum was used in American breweries in the 1940s when traditional ingredients were scarce due to the war, but quality problems led to it being abandoned.

Unmalted wheat: Unmalted wheat is used in some recipes that require its specific attributes; namely, the raw grain flavor and cloudy appearance associated with Belgian white beer. The gelatinization temperature is lower than barley, so it can be mixed into the mash directly.

Oats: Oats are low in starch, high in oil and protein, and extremely high in beta-glucans. As a result, they are not used as a major substitute for malt in the grist. However, they add a smoothness and increased mouthfeel to beers and are popular as an additive to stouts.

Rye: Rye is another grain that is used for its flavor, rather than as a malt replacement. It has a strong distinctive bite but is difficult to lauter, so extract recovery is difficult and slow. Rye is known to impart an orange tinge and a spicy character to beer.

BREWING WITH MALTED GRAINS

Adjuncts such as wheat, rye, and oats are all available as malted grains, though they are often not quite as easy to brew with as malted barley. They should be used for specific flavor or quality

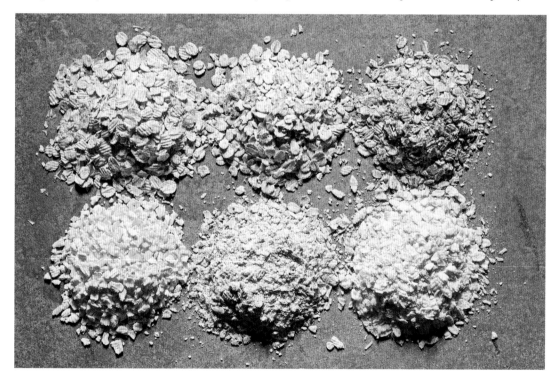

Adjuncts such as oats and corn are typically added to the mash.

contributions. To use malted adjuncts, all-grain brewers simply crush the grains along with the malted barley and add them to the mash. Extract brewers must do a partial mash. Malted adjuncts contain amylase enzymes, so they can convert their own starch into sugars without the addition of malted barley to the partial mash (see page 24 for instructions on performing a partial mash).

Malted wheat may be 50 to 75 percent of the grist in a German wheat beer. Since wheat has no husk, brewing with malted wheat can be tough. Many brewers use rice hulls to establish a good filter bed in the mash. Rice hulls replicate the role of barley husk in the lautering stage and hence aid wort separation. Wheat adds a tangy character and improves head formation and retention.

Rye malt makes up to 10 percent of the grain bill in rye beers. Rye cannot be used in larger quantities because it contributes to stuck mashes.

BREWING WITH FLAKED GRAINS

All-grain brewers can simply add flaked grains to their mash. Although flaked grains have no amylase enzymes, excess enzymes from the barley can degrade the starch. As with malted grains, extract brewers must perform a partial mash if they wish to use flaked grains. When mashing these flaked grains, the brewer must also add barley malt to supply enzymes. A 1:1 mixture of flaked grains and 6-row barley malt is usually sufficient. Crush the barley malt and place it in a grain bag with the flaked grain. The flaked grain does not need to be crushed, but it helps to break it up. Once the grains are mixed, steep the grain in 150 to 158°F (65 to 70°C) water for 30 minutes, rinse the grains, and proceed.

HOPS

The next time you open a beer, fill your glass halfway. Give the beer a good sniff; get your nose way down in there like the wine snobs do and enjoy the aroma. Does it smell flowery? Fruity? Pine scented? Now take a drink, just enough to fill your mouth. Consider how the beer tastes on the front of your tongue and on the back of your tongue. Is it bitter? Spicy? Grassy? These are all qualities, both good and bad, that hops can add to beer. This versatile climbing vine works in several ways to add widely differing characteristics to your brew.

HOP BASICS

Historically speaking, hops are a fairly recent innovation in the brewing world. Although evidence exists of their cultivation as early as A.D. 200 in Babylon, and A.D. 700 in Germany, they were not widely used in brewing until the eleventh century in Bavaria. They didn't gain wide acceptance until the fifteenth or sixteenth century in the rest of Europe. In England they were not highly thought of initially, and King Henry VIII banned their use. This was only a short time before brewers began emigrating to the United States, so American brewers have been using hops for about as long as their European counterparts.

It is impossible to think of a beer today that does not include hops. In fact, it is a legal requirement in the United Kingdom that beer includes hops in the formulation. The Bavarian purity law, the Reinheitsgebot, written in 1516, also legislates the use of hops for German brewers.

Hops provide bitterness, a nice flavor, and a pleasant aroma to beer. They enhance the foam on beer and the way the foam clings to the side of the glass. They also provide protection against beer spoilage from certain microorganisms. Over the centuries of hop use, microorganisms have evolved that are resistant to hops, but most of these are found only in breweries.

Now for the favorite ingredient of many brewers: hops!

Hop cones used for brewing are the dried seed cases of the hop plant *Humulus lupulus*. The hop plant is a vining plant that can grow as tall as 30 feet (9 m) high. The vines of the hop plant are called bines, because they grab their support surfaces with a multitude of tiny hairs, not the grasping tendrils of a true vine. The hop plant produces flowers that are small, green, and spiky, somewhat resembling a burr. From these flowers the hop cone (or strobile) is formed. Hop cones consist of a central strig or stalk, and between twenty and fifty petals, called bracts. At the base of these bracts, the resin (known as lupulin) is produced as a sticky yellow powder.

The hop plant is a perennial with separate male and female plants. All commercial hops, used for flavoring beers of all sorts, grow on the female plants and will contain seeds if male plants are allowed to produce pollen near them. To prevent seeds from developing in the hops, most countries do not permit male plants to be grown anywhere. However, in England, male hops are permitted (except in Hampshire) and most English hops contain seeds.

The plants grow up strings or trellis wires during the summer, and the hops are harvested and dried in September. In England, the United States, Canada, and Australia, hops are usually packed into the final package on the farm where they were grown. In European countries, the individual farmer's hops are usually blended, redried, and packed into bales in large lots or processed directly from the farmer's lots.

There is a great deal of variation between hops from the different countries, as well as different growing regions within a country—and even from farm to farm. Typically a hop cone consists of the following components: 10 percent water, 15 percent total resins, 0.5 percent essential oil, 4 percent tannins, 2 percent monosaccharides, 2 percent pectins, 0.1 percent amino acid, 3 percent lipids and wax, 15 percent proteins, 8 percent ash, and 40.4 percent residual carbohydrate (cellulose, lignin).

Brewers are largely interested in the total resins and the essential oils, which represent the brewing value of the hop. Both are contained in the yellow lupulin dust that is found around the base of each bract on the hop cone. This material is essentially the only portion of the hop a brewer need be concerned with. The rest of the hop's plant matter may perform an important role in the brewery as a separation aid. The green material acts as a filter screen, which aids in clarifying the wort after it has been boiled. The other components, particularly proteins and polyphenols, are soluble in boiling wort, although it should be remembered that greater quantities of protein and polyphenols are derived from malt.

The total resins arc further subdivided into hard resins, soft resins, and uncharacterized soft resins. Soft resins consist of alpha and beta acids, and it is those compounds that the brewer is most interested in.

ALPHA ACIDS

Alpha acids comprise more than 50 percent of the soft resins and are largely thought of as the primary source of bitterness in beer. Not directly, though, as they are insoluble in wort and must first be isomerized by heat to become soluble. It requires around 45 minutes of boiling to isomerize and solubilize 30 percent of the potential alpha acids from the hops. This amount drops dramatically as the boiling time diminishes. The isomerization reaction results in a change in the chemical structure of the alpha acid molecule.

Alpha acids belong to a class of compounds known as humulones. They consist of a complex hexagonal molecule with several side chains, including ketone and alcohol groups. Examples of humulones include humulone, cohumulone, adhumulone, posthumulone, and prehumulone. Each different humulone differs in the makeup of the side chain. For instance, humulone has a side chain of isovalerate attached, while cohumulone has isobutyrate as its side chain. These side chains can become detached during extended storage under poor conditions and result in the cheesy flavors sometimes associated with old hops. It has become accepted dogma among brewers to think of each of these humulones as having different bittering characteristics. Some swear that the bitterness associated with cohumulone is harsher than that from humulone. Other studies have shown no difference in sensory impact when each of the different humulones is compared. Nevertheless, the humulone-to-cohumulone ratio is now quoted in hop analyses and new varieties are being bred with low cohumulone levels in mind. Historically, the most highly prized hop varieties—including noble hops such as Hallertau, Tettnang, and Saaz—also happen to be those that have low cohumulone levels.

The alpha-acid levels in hops begin to tail off immediately after harvesting and continue to decline in storage. The number quoted to you on a packet of hops is the alpha-acid content when the hops were tested immediately after harvest. Despite the best intentions of the retailer, the hops have been subjected to conditions that cause their alpha-acid levels to be lower. High temperature and exposure to air will speed up the losses of alpha acids. In hop varieties with poor storage characteristics, up to 50 percent of the total harvest alpha acids may be lost in 6 months when hops are stored at 70°F (21°C). A good hop will still lose 20 percent of its total acids under the same storage conditions. Hops should be stored in a fridge, or preferably a freezer, and air must be excluded from the package. Since you have no way of knowing what the hops experienced before you bought them—remember, the inside of a truck can get up to 140°F (9°C) in the summer in some southern states—it is always better to buy your hops from a reputable hop supplier.

Additional benefits of alpha acids are seen from their role in foam formation and head retention. They cross link chemically with certain specific proteins in an extremely complex manner to support foam. If you sip the thick foam on a pint of nitrogen-poured Guinness, you will notice a distinctly more bitter taste than that found in the beer beneath.

BETA ACIDS

These compounds are not actually bitter but will turn bitter when they oxidize during storage. The alpha-to-beta ratio is considered important in gauging how a hop will provide bitterness as the hops age. The bittering potential from alpha acids declines with time, but the bittering potential from oxidized beta-acids increases. In a hop with a 2:1 ratio of alpha to beta acids, the bittering potential may remain fairly constant. The oxidation

Vacuum-sealed packaging is a good indicator that the hops you ordered have been well taken care of.

reaction will take place to an even greater extent during kettle boiling. Beta acids consist of lupulone, colupulone, adlupulone, and other substances and, like alpha acids, differ in the structure of the side chains. Again there is a difference of opinion in the brewing world as to the character of bitterness derived from beta acids compared to that of alpha acids. In Germany, oxidized beta-acid bitterness is preferred while in Japan it is considered too harsh.

OILS

The total oils, formed in the lupulin glands, represent the general aroma characteristics of the hop. It varies in concentration depending on hop variety and from season to season. It may be as low as 0.5 percent or as high as 2 percent. The oils are soluble in boiling wort but are extremely volatile and are largely lost during the wort-boiling phase of brewing. A full boil of an hour to an hour and a half, needed to volatalize most of the unfavorable aroma characteristics from the malt and precipitate enough of the denatured protein and polyphenols, results in the complete loss of any of the aroma components from the hops. Brewers get around this issue by adding a portion of the hop charge to the boiling wort 5 to 10 minutes from the end of the boil. Alternatively, brewers add hops immediately after boiling, but before chilling, to attempt to extract the aromas and avoid the losses due to volatilization.

The action of yeast fermenting sugar and causing vast amounts of CO_2 to rise through the wort has the effect of carrying hop aroma away with it. While this may produce a wonderful hop aroma in your fermentation area, it will cause a decrease of hop aroma in the beer. Remember that intense hop aromas are not always pleasant or desired depending on the beer style, so losing some may be a good idea. Hop aroma may also be added to the finished beer by a process known as dry hopping, in which whole hops are added to the beer in a maturation vessel. For more on dry hopping and a deeper discussion of hop oil components, see page 92.

BUYING AND STORING HOPS

There are many different varieties of hops available to the homebrewer. In addition, hops come in a couple different forms, which vary with regards to their storage potential and performance in the brewhouse.

Whole hops are simply hop cones that have been picked and dried in large kilns called "oasts." Some European hops may also have been sulfured, meaning sulfur sticks were burned in the kiln to give the hops a uniform green appearance. German hops are almost always sulfured. In contrast, no hops grown in the United States are sulfured. Whole hops, or cone hops, are sometimes called leaf hops or whole leaf hops, although the leaves of the hop plant are not used in brewing.

Whole hops are only lightly processed (just picked and dried) and this appeals to some brewers. Many prefer to use whole hops when dry hopping—even if they use pellet hops in the kettle—as they feel their aroma qualities are better. In storage, the lupulin glands of whole hops are exposed and will oxidize faster than pellet hops when exposed to oxygen. Under optimal storage conditions—frozen, inside an oxygen-barrier bag—the difference is minimal.

When boiled, whole hops float on top of the wort, and it is easy to siphon clear wort out from underneath them after the boil. When brewing with a kettle with a spigot at the bottom, a screen at the bottom of the kettle may be employed to keep the hops from exiting the kettle. Whole hops may be used in a hopback (or a Randall) as the cones will form a filter bed for the wort or beer. In the same situation, pellets would dissolve into a sludge, blocking the flow of the liquid.

Many brewers love fresh, whole cone hops. There's nothing like adding handfuls to an IPA.

Note: *Many homebrewers grow their own hops and hence have whole hops to brew with. However, measuring the alpha acid level in them requires laboratory analysis and equipment beyond that found in a typical homebrewery. Hop producers, larger breweries, and some independent laboratories routinely measure the alpha acid level of hops. For a fee, it is possible to send samples to these labs for precise analysis.*

Pellet hops are whole hops that have been finely milled and extruded through a die. Pellets are widely used by homebrewers and commercial brewers alike because they are much more compact than whole hops and have better storage characteristics when storage conditions are not optimal. Pellet hops also show about 10 percent better hop utilization than whole hops.

The friction from pressing the hop material through the die causes it to heat up. In the past, this heat led to the degradation of some volatile hop components. These days, pellet dies are cooled by liquid nitrogen and the flow rate through the die and temperature of the extruded product are monitored to keep these losses to a minimum.

Created by Yakima Chief Hops, Cryo hops have become a popular option for brewers looking to maximize hop flavor and aromas while minimizing beer loss. These pellets are made up of the concentrated lupulin of whole-leaf hops through a process of cryogenic freezing. The process separates the lupulin, containing resins and aromatic oils, from the leafy material (bract). Because of the high oil concentration brewers can use less hops (about half the amount of traditional T-90 pellets), resulting in less trub loss. Cryo hops are also said to reduce the astringent flavors of

Pellet hops are designed to be more efficient than whole cone hops.

vegetative material that sometimes comes with using large hop additions in the whirlpool or dry hop in styles like New England IPA.

In the kettle, pellets dissolve and the material from them sinks to the bottom of the vessel. (They also sink when used as dry hops.) As such, whirlpooling your wort—stirring it in a circular motion, then letting the debris settle in the middle of the kettle—is advised before racking the wort to your fermenter.

CO_2-hop extract is the extracted hop resins and oils from hop pellets. Extracts are generally only used on the hot side as a bittering charge, where there are many benefits to using them over pellets. Hop extracts provide an economical bang for your buck in increased yields from reduction of solids, reduction of trub formation, and reduction of foam. Extracts can also produce a cleaner, brighter beer while maintaining hop varietal impact. The CO_2 extraction process also reduces and/or eliminates the agrochemical (pesticide) and nitrate content of hops. Another benefit is that extracts can be stored at ambient temperature for years and years. On the homebrew scale, extracts are generally sold by the syringe at a dosage appropriate for 5-gallon (19 L) batches.

EVALUATING HOPS

When buying hops, you should search out those that look green, not brown. In addition, hops should smell fresh, not cheesy. A standard way of evaluating hop aroma is to take whole hops in your hands, rub them together briskly three or four times, then quickly put your nose right down in the sample and smell it. Rubbing ruptures the lupulin glands and releases aroma.

At your homebrew shop, hops should be stored frozen and packaged in oxygen barrier bags. Hop resins are somewhat unstable and prone to oxidation, and the alpha acid level of hops decreases over time. This loss is decreased at lower temperatures and when hops are shielded from exposure to oxygen. The storage stability also varies with the variety of hops. Some varieties will lose more than 50 percent of alpha acids if stored for a year at room temperature; the loss is less than 15 percent if stored at 0°F (-18°C). Of course, you will not be able to smell the aroma of frozen and properly bagged hops until you open the package.

Hops should be stored in your freezer at home until it is time to use them. If you are going to use the hops within a few months, an ordinary frost-free freezer—like that found in almost all home refrigerators—is fine. (For short-term storage, up to a couple weeks, your refrigerator is fine.)

For long-term storage, hops should be placed in a freezer that does not go through freeze-thaw cycles if at all possible. If you are storing your hops for 6 months or longer, they would be better off permanently frozen. In this state, they can last for years. The alpha acids will still degrade over time, but the hops will still smell good and be suitable for brewing.

Some homebrewers buy their hops in bulk late in the year after the harvest comes in. Big packages of hops should be broken down into smaller packages once opened. Regular plastic zipper bags are oxygen permeable. If you're stuck with this as your only option, double bag your hops to minimize exposure to oxygen. Food sealers, such as Seal-A-Meal or Food Saver, are great for repackaging bulk hops into smaller portions. These devices remove most of the air (and hence the oxygen) from the portion, then seal the bag. The bags themselves can be stored in a large canning jar. If you plan to store hops for any length of time, be sure to label the bag with the variety, the alpha acid rating, and the date you purchased them.

HOP VARIETIES

When you start investigating homebrewing ingredients, it's easy to feel a little overwhelmed by the number of different hop varieties to choose from. Thankfully, many brewers have come before you and have figured out which hops work with each beer style. And of course once you get a handle on the classics, the sky is the limit for experimenting with different hop varieties in your homebrewery. On the pages that follow, you'll find a rundown of some of the more common beer style/hop pairings and a chart of useful information for many of the hop varieties that are available to homebrewers.

Try growing fresh hops if you can manage it in your growing region.

For an up-to-date listing of hops that are available to homebrewers, check out BYO's online chart at www.byo.com/resource/hops.

American pale ale, IPA, DIPA, and barleywine: If you're interested in brewing recipes in this category, start your research with a classic Sierra Nevada pale ale (see recipe on page 224). Pale ales range from about 25 IBUs all the way up into the 60-plus IBU range these days thanks to hop heads, and it's not unheard of to hit 100 IBUs for IPA, DIPA, and barleywine. Since it's got "American" right in the name, the key to choosing hops for these ales is to stick with varieties that originate in the United States. For example, Sierra Nevada's pale ale leans heavily on Cascade. You could also go with Chinook, Centennial, Columbus, Simcoe, or Amarillo. If you want to pull back on the big, aggressive hoppyness, you could also use Willamette, which is a clone of Fuggle, or you could try Mount Hood, which is a clone of Hallertauer Mittelfrüh, one of the classic noble hops. Newer and trendier hop varieties these days often focus on unique aromatic properties. For example, it wasn't so long ago that Citra and Mosaic were introduced. They are now very popular hops for pale ales and IPAs.

English pale ale, IPA, barleywine: The times are changing everywhere in craft brewing. Some of the newer British brewers are pumping up the IBUs in their ales to keep up with craft beer drinker's thirst for hops. Classically, however, English pale ales and IPAs are more restrained than their American counterparts and range between 25 to 35 IBUs. Overbittering is not appropriate for this style, and of course they are brewed with traditionally English hop varieties such as Fuggles, East Kent Goldings, and Target. The same rules apply as you push into the more highly hopped IPAs and barleywines—keep restraint in mind, and be sure the hops are British.

Porter, stout, brown ale: The IBUs of these three styles can clock in all over the map, and the heavy-handed hop creep that started in pale ales and IPAs has found its way into these categories as well. Unlike pale ales, however, most standard bittering hops are just fine for the boil additions—the flavors of the roasted malts will overpower the boil hop aromas, so just choose something that will give the beer a good backbone, such as Cluster, Magnum, or Galena. For aroma hops, try Northern Brewer, Fuggles, or Willamette.

American lager: This category has long been dominated by macrobrewed pale lagers such as Budweiser and Coors. However, hoppy lagers are increasingly popular. Like the other aforementioned categories, the IBUs in this category can range from as low as 10 to 12 for the macrobrewed pale light lagers all the way up to the mid-60s and even 70s for India pale lagers such as Jack's Abby Hoponius Union (see recipe on page 200). If you want to brew like the big guys (AB-InBev, Miller, Coors, etc.), lean toward low-alpha varieties that are very neutral, such as Hallertauer, Liberty, Mount Hood, and US Saaz, to create a subtle hop flavor. If you're going down the IPL road, take more of an IPA approach to choosing hops—big, bold, aromatic American varieties work great, such as Cascade, Chinook, and Columbus. As long as you're breaking some style rules, however, you could also try using a blend of English and American varieties, such as Willamette, Glacier, Cascade, Target, and Fuggle.

German ale and lager: This category includes Märzen, Vienna, Dortmunder export, dunkel, and bock for lagers, and Kölsch and altbier for ales. (Pilsners and wheat beers will be discussed in a moment.) German beers place more of a premium on maltiness than bitterness and have been successfully brewed with traditional German noble varieties of hops for hundreds of years, so there's really no guesswork needed here. Choose Hallertauer Mittelfrüh, Hallertau Tradition, Spalt, Hersbrucker, Perle, Saphir, Select, or Tettnanger for any of these beer styles.

European Pilsner: Your choice for hopping a European Pilsner will obviously come down to the origin of the style you are brewing—Czech or German. These beers traditionally clock in at around 25 IBUs, and you have your choice of hop varieties that come from either the Czech Republic (Czech Saaz, Premiant, Sladek) or Germany (Tettnanger, Hallertauer Mittelfrüh, Spalt).

Wheat beer: Wheat beers tend to be more yeast-driven and as such should be hopped with restraint. As with all beers mentioned earlier, wheat beers are not immune to hop creep, and some modern examples of American wheat beer can be pretty hoppy. However, most wheat beers hover somewhere between 10 and 30 IBUs. Choose hops that match the origin of the beer you want to brew (American varieties for American wheats, German varieties for your hefeweizens and weizenbocks, and so on).

Hop Name	Country	Alpha Acid	Possible Substitutions	Flavor Description
African Queen	South Africa	10–17%	Amarillo, Cascade, Mosaic	Gooseberries, melon, cassis, chilies, gazpacho
Ahtanum	US	4–6.3%	Cascade, Willamette	Citrus (grapefruit, lemon), geranium, floral, pine, earthy notes
Altus	US	15–18.5%	Apollo	Massive resinous, spicy, tangerine aromas
Amarillo	US	8–11%	Cascade, Centennial	Citrus (orange, grapefruit, lemon), peach
Apollo	US	15–19%	Nugget, Columbus	Lime, pine, resin
Azacca	US	14–16%	Citra, Mosaic, Galaxy	Apricot, ripe mango, orange, grapefruit, pine, pineapple
Barbe Rouge	France	7–10%	Monroe	Red ripe fruit flavors of currant, strawberry, raspberry
Bramling Cross	UK	5–7%	UK Kent Golding, UK Progress, Whitbread Golding Variety	Grapefruit, vanilla, gooseberry
Bravo	US	14–17%	Apollo, Zeus	Orange, fruity, vanilla, floral
Brewer's Gold	UK, German	6–7%	Bullion, Galena, Northdown, Northern Brewer	Blackcurrant, fruity, spicy
BRU-1	US	13–15%	Sultana, El Dorado	Pineapple, stone fruit, spice, floral
Calypso	US	12–14%	Galena, Huell Melon	Apple, pear, lime, fruit punch
Cascade	US	4.5–7%	Amarillo, Centennial, Columbus	Pleasant, flowery, spicy, citrusy. Can have a grapefruit flavor
Cashmere	US	7–10%	Cascade, Lemondrop	Lemon, lime, melon, subtle herbs
Centennial	US	9.5–11.5%	Chinook, Columbus	Floral, lemon, orange
Challenger	UK	6.5–8.5%	Northern Brewer, US or German Perle	Spicy, cedar, green tea
Chinook	US	12–14%	Galena, Nugget, CTZ	Spicy, piney, grapefruity
Citra	US	10–15%	Simcoe, Idaho 7	Citrus, grapefruit, melon, gooseberry
Cluster	US	5.5–8.5%	Galena	Spicy, earthy, floral
CTZ (Columbus, Tomahawk/Zeus)	US	14–18%	Chinook, Northern Brewer, Nugget	Earthy, spicy
Comet	US	8–12%	Galena	Grapefruit, lemon, orange, grassy notes
Contessa	US	3–5%	Nobel hops	Herbal, green tea, floral, light pear
Crystal	US	2–6%	French Strisslespalt, Hallertauer, Hersbrucker, Liberty, Mt. Hood	Woods, floral, herbs, spice
East Kent Golding	UK	4–6.5%	UK Progress, Golding, Whitbread Golding Variety	Lavender, honey, earth, lemon, thyme

Hop Name	Country	Alpha Acid	Possible Substitutions	Flavor Description
Ekuanot	US	14.5–15.5%	El Dorado, Huell Melon	Melon, citrus, apple, mango
El Dorado	US	13–17%	Galena, Ekuanot	Tropical fruit, watermelon, pear, mango
First Gold	UK	6–10%	UK Kent Golding, Styrian Golding	Geranium, apricot, tangerine
Fuggle	UK	3–5.5%	Styrian Golding, US Fuggle, Willamette	Great tea, mint, gass, floral
Galaxy	Australia	11–16%	Amarillo, Citra	Passion fruit, peach, citrus
Galena	US	10–14%	Chinook, Nugget, Pride of Ringwood	Medium but pleasant hoppiness, citrusy.
Glacier	US	5–9%	Nugget, CTZ	Pear, pineapple, lime, black currant, grapefruit
Golding	US	4–6%	East Kent Golding, UK Progress, Whitbread Golding Variety	Mild spice, herbal
Hallertauer	US	3.5–5.5%	Hallertauer Tradition, Liberty, Ultra	Mild spice, floral
Hallertau Blanc	German	9–12%	Nelson Sauvin	Pineapple, white grape, lemongrass, passion fruit
Hallertau Mittelfrüh	German	3–5.5%	Hallertau (US), Liberty, Ultra	Fresh cut hay, sweet spices, tea
Hallertau Tradition	German	4–7%	Crystal, Liberty	Grass, tea, earthy, orange
Hersbrucker	German	1.5–5%	Strisslespalt, Mt. Hood	Hay, tobacco, orange
Horizon	US	11–14%	Magnum	Spicy, floral, citrus
Huell Melon	German	7–8%	Ekuanot, Cascade	Honeydew melon, strawberry aroma, tropical fruit
Idaho 7	US	13–15%	Mosaic, Citra	Citrus (orange, grapefruit, papaya), apricot pine, resin, black tea
Kohatu	New Zealand	6–7%	Motueka	Woodsy, pine needles, tropical fruit
Lemondrop	US	5–7%	Cascade, Cashmere, Centennial	Lemon, citrus, herbal, mint, green tea
Loral	US	10–14%	Strisselspalt, Nugget	Dark fruit, floral, lemon, citrus, pepper
Lotus	US	13–17	Bravo	Orange, vanilla, candied grapes, tropical fruit
Magnum	German	11–16%	Hallertauer Taurus, Nugget	Apple, pepper, citrus, clean bittering
Mandarina Bavaria	German	7–10%	Columbus, Nugget	Mandarin orange, tangerine, lime
Medusa	US	3.5–5%	Zappa	Guava, citrus, melon, apricot, woody
Mosaic	US	11.5–13.5%	Citra, Nugget, Simcoe	Blueberry, tropical fruit, citrus, grassy, pine, floral

Hop Name	Country	Alpha Acid	Possible Substitutions	Flavor Description
Motueka	New Zealand	6.5-7.5%	Saaz, Sterling, Liberty	Lime, lemon, tropical fruit
Mt. Hood	US	3-8%	Crystal, French Strisslespalt, Hersbrucker	Nobel-like, herbal notes, flowers, green fruit
Nelson Sauvin	New Zealand	12-14%	Pacific Jade, Hallertau Blanc	Sauvignon blanc wine, gooseberry
Northdown	US	6-10%	Challenger	Spice, floral, pine, cedar, berry
Northern Brewer	German, UK, US	6-10%	Chinook, Perle	Mint, pine, grass
Nugget	US	9.5-16%	Columbus, Magnum	Woody, pine
Pacific Gem	New Zealand	13-16%	Galena, Pacific Jade	Blackberry, black pepper, herbal
Pacific Jade	New Zealand	12-14%	Galena, Pacific Gem	Blackberry, citrus, floral, black pepper, oak
Pacifica	New Zealand	12-14%	Hallertau Mittlefrüh, Perle	Orange zest marmalade, floral
Pahto	US	17-20%	East Kent Golding	Herbal, earthy, floral
Palisade	US	5.5-10	Willamette	Apricot, grass, floral
Pekko	US	13-16%	Heull Melon, El Dorado	Floral mint, melon, thyme, sage, cucumber
Perle	German, US	4-9%	Northern Brewer	Spicy, floral, tea
Phoenix	UK	8-13%	Northern Brewer, Challenger, Northdown	Spicy, floral, chocolate
Pilgrim	UK	9-13%	UK Challenger, UK Kent Golding, UK Northdown	Pine, chocolate, molasses, floral
Pioneer	UK	9-13%	Challenger, East Kent Golding	Lemon, grapefruit, spicy, herbal
Polaris	German	8-10%	Northern Brewer	Mint, pineapple, methol
Pride of Ringwood	Australia	7-10%	Cluster, Centennial	Woody, spicy, herbal
Progress	UK	18-24%	Fuggle, East Kent Golding	Mint, grass, honey
Rakau	New Zealand	5-11%	Summit, Amarillo	Stone fruit, apricot, fig, pine
Riwaka	New Zealand	9-11%	Saaz, Motueka	Grapefruit, tropical fruit, citrus
Saaz	Czech	2.5-4.5%	US Saaz, Sterling	Mild earthy, herbal, floral
Sabro	US	12-16%	El Dorado, Southern Tropic	Tangerine, coconut, tropical and stone fruit, mint
Santiam	US	5-7%	Spalt, Tettnang	Mild black pepper, spicy, herbal
Saphir	German	2-4.5%	German Spalt, Hallertau Mittlefrüh	Spicy, fruity, floral
Simcoe	US	11-15%	Summit, Citra, Mosaic	Passion fruit, citrus, berry, bubble gum, earthy

Hop Name	Country	Alpha Acid	Possible Substitutions	Flavor Description
Sorachi Ace	Japan	11–16%	Liberty, Lemondrop, Southern Cross	Lemon, dill, orange
Southern Cross	New Zealand	11–16%	Simcoe, Lemondrop	Lemon, lime, pine
Southern Passion	South Africa	5–12%	Southern Tropic	Passion fruit, guava, coconut, citrus, red berries
Southern Star	South Africa	12–18.6%	Mosaic, Ekuanot	Pineapple, blueberries, tangerine, pine resin, herbal spice
Sovereign	UK	4.5–6.5%	Fuggle	Floral, herbal, grass, pear
Spalt	German	4–5.5%	Saaz, Hallertau	Mild, slightly spicy, earthy
Sterling	US	4.5–9%	Saaz, Mt. Hood	Mild, herbal, slightly spicy
Strisslespalt	France	1–5%	Mt. Hood, Crystal	Spicy, citrus, floral, herbal
Styrian Golding	Slovenia	4.5–6%	Fuggle, Willamette	Resinous, earthy
Sultana	US	13–15%	Nugget	Pineapple, citrus, pine
Summer	Australia	5.6–6.4%	Belma, Palisade	Delicate floral, slightly earthy, apricot, stone fruit
Summit	US	16–18%	Columbus, Simcoe	Pepper, incense, citrus
Super Pride	Australia	12.5–16%	Pride of Ringwood	Resin, fruit
Target	UK	8–13%	Fuggle, Willamette	Citrus, sagebrush, spice
Tettnanger	German	2.5–5.5%	Saaz, Spalt	Fresh herbs, pepper, dried flowers, black tea
Topaz	Australia	13.7–17.7%	Summit, Apollo	Resinous, grassy, lychee
Triumph	US	9–12%	Hallertauer Mittelfrüh, Hersbrucker	Peach, lime, orange, spice, pine
Vanguard	US	4–6%	Hallertauer Mittelfrüh, Saaz	Wood, earthy
Vic Secret	Australia	14–17%	Galaxy	Resinous, pine, pineapple
Wai-iti	New Zealand	2.5–3.5%	Kahatu	Mandarin orange, peach, lime zest
Waimea	New Zealand	16–19%	Pacific Jade, Pacific Gem	Citrus, pine needles
Wakatu	New Zealand	6.5–8.5%	Hallertauer Mittelfrüh	Floral, lime zest, citrus
Warrior	US	15–18%	Nugget	Resin, pine, citrus
Willamette	US	4–6%	Glacier, Styrian Golding, East Kent Golding	Floral, incense, elderberry
Zappa	US	6–8%	Medusa	Spicy, mango, passion fruit, citrus, pine

HOPPING METHODS

Hops are added at different stages during the brew day to achieve different effects. Most homebrew recipes will indicate when to add the hops: early for bittering, mid to late for flavor, and at the end of the boil and in the fermenter for aroma. Bittering hops are added when the wort is first collected in the brewpot, either just before the boil (first wort hopping) or when it starts to boil. Flavor hops are typically added when there are 15 to 30 minutes left in the boil, which will provide some additional bitterness, but the hops will not boil long enough to allow the heat to drive off the hop flavor. Finally, aroma hops are added either at the end of the boil, at "flameout" (when the heat is turned off), in the whirlpool, or in the fermenter as dry hops.

FIRST WORT HOPPING

As far as adjusting a brew day technique, it really doesn't get any easier than trying first wort hopping (FWH). While there is still some ambiguity that surrounds its overall affect on the finished beer, most brewers will either use it as a substitute for their bittering hop charge or their mid-boil hop charge. So why the ambiguity? It has been shown in tastings that FWH can increase the perceived aroma of a beer when substituted for a traditional late hop addition to the kettle. Yet when a group of scientists tested the hop aroma components of FWH beers, they found that the aroma compounds were actually considerably lower in FWH beers versus the same beer with that late hop addition. So let's delve into FWH to find out more.

FWH is very simple: add a portion or all of your late boil hop charge or bittering hops before the wort comes to a boil. It does not matter whether you are an all-grain or extract brewer. Generally as an all-grain brewer you'd do this about 3 to 5 minutes into the sparging process of the grain bed. When you're using extracts, you'd generally wait until the wort has gotten up to about 180°F (82°C) before tossing the FWH into the kettle.

So what does FWH do for your beer? What repeated studies have shown from blind triangle taste tests is that it creates a softer, more rounded bitterness than adding your bittering hops to a rolling boil. Two studies in particular document this effect. The first is a fairly comprehensive study put out back in 1995 by a German group of researchers, Preis, Nuremberg, Mitter, and Steiner, titled "The Rediscovery of First Wort Hopping," published by Brauwelt International. The second study was performed by US homebrew guru Denny Conn, whose results roughly affirmed those of the German researchers, and he presented his findings at the 2008 National Homebrewers Conference.

The German researchers utilized two breweries to test FWH versus a late hop addition. The taste tests from both breweries confirmed that there is a distinct difference between a beer with first wort hopped charge and bittering charge and beers brewed with a traditional bittering charge and late hop charge. Among those on the panel, twenty-one out of twenty-three tasters were able to recognize the taste discrepancy in a Pilsner. Of those twenty-one tasters who distinguished some difference between the two beers, nineteen of them preferred the first wort hopped beer. Denny Conn's study, which was performed with two groups of Beer Judge Certification Program judges and professional brewers, tested FWH beers versus a traditional bittering charge. What they found in a blind triangle test was that seven of the eighteen tasters were able to distinguish the first wort hopped beer—still significant but not quite as striking as the German study. The general consensus among those who could distinguish among the two beer types was that the first wort hopped beer offered a smoother bittering profile than the reference beer.

The other aspect that the two studies confirmed is that first wort hopping will increase your bittering units without increasing the perceived bitterness. The IBUs of a beer that has been first wort hopped achieve on average about 10 percent more hop utilization when analyzed against the same beer with a standard 60-minute hop addition. The Germans found in the first Pilsner FWH 39.6 IBUs, compared with 37.9 IBUs for the reference beer. The second Pilsner was found to have 32.8 IBUs, and the reference beer had 27.2 IBUs. Denny Conn's beers, when analyzed, had 24.8 IBUs for the FWH beer and 21.8 IBUs for the reference beer. The German study also showed that the iso-alpha acid concentration was quite a bit higher for the FWH beers. But again, taste tests revealed that the actual perception of the bitterness levels are slightly lower for the FWH beers. In other words, they had less bite.

So when is it advisable to utilize the FWH technique? It's up to your personal preference, but some styles that are a good fit for a trial are malt-forward ales and lagers or well-rounded styles such as continental lagers, roasted grain-focused beers, wheat beers, and Scottish ales. On the other hand, you may want to skip FWH and stick to a more traditional hopping schedule when you're looking for more bite in the beer, such as with American IPAs, double IPAs, robust porters, or imperial stouts. Of course, you could also try splitting the bittering hops so that half go in at first wort and the second half go in with 60 minutes left in the boil.

Note: *Mash hopping is another preboil hopping technique, but it is not one used by many brewers. Unlike FWH, iso-alpha acid conversion does not take place at an appreciable level at mash temperatures, so alpha acids do not undergo the isomerization reaction. Furthermore, the oils that are extracted from the hops will most all be driven off during the boil.*

HOP STANDS, WHIRLPOOL HOPPING, AND HOP BURSTING

A *hop stand* is simply allowing boiled wort an extended contact period with the flameout hops prior to chilling. Pro brewers typically create a whirlpool either in their kettle or in a separate whirlpool vessel with the hot wort and the ensuing vortex creates a cone-shaped pile in the center of the vessel made up of the unwanted trub and leftover hop material, hence many pro breweries will refer to this technique as *whirlpool hopping*. (As a homebrewer working with a much smaller volume, it's not necessary to keep the wort constantly swirling in the kettle for this technique to work, though.)

Whirlpool hopping is a great way to add unique hop aroma and flavor to your beer.

Another slight variation on this technique is adding all of your boil's hop additions—including the hops for bittering—at this point. This is known among homebrewers as *hop bursting*.

In short, all of these techniques allow the hops added at flameout a period to release their essential oils into the wort, while minimizing the vaporization of these essential oils. In essence, it adds a kick of hop flavor and aroma while also adding what can best be described as a smooth bitterness. Read on to find out more about how these techniques can add significantly to the hop flavor and aroma of beer.

A WORD ON ESSENTIAL OILS

The essential oils found in hops are volatile and provide beer with the hop flavor and aroma hop aficionados enjoy. While there are hundreds of essential oil components, for practical purposes brewers tend to focus on four to eight main essential oils that play vital roles in providing hop varietal characteristics. One important characteristic is the essential oil's flashpoint, or the temperature at which the essential oil is actively vaporizing to the point where it could ignite if sufficient vapors were present. At wort boiling temperatures, all hop essential oils have surpassed their flashpoints, so a vigorous boil will drive them off fairly quickly.

The best way to think about the driving-off process is in terms of half lives. The lower the flashpoint, the faster the oil vaporizes and the faster the half life. The longer the hops are boiled and the lower the flashpoint, the less the essential oil will impact the beer. In effect, whirlpool hopping removes the rolling boil (for the whirlpool hops), lowering the temperature of the wort and therefore reducing the vaporization rate of the essential oils, allowing the essential oils to really soak in to the wort.

ALPHA ACID ISOMERIZATION

Alpha acids will continue to isomerize after flameout until the temperature of the wort reaches about 175°F (79°C). Homebrewers trying to calculate a beer's IBUs will need to guesstimate how much isomerization is occurring. The closer the wort is to 212°F (100°C), the higher the alpha acid isomerization rate. To do this, we can look to professional brewers for some guidelines. Ultimately, however, the thermal capacity of a professional 60 bbl whirlpool vessel is quite different than 5 gallons (19 L) of homebrew, so the comparisons can only be rough guidelines at best.

Matt Brynildson of Firestone Walker Brewing Company says, "The fact that there is some isomerization [about 15 percent in whirlpool versus 35 percent in the kettle] of alpha acid means that not only hop aroma and hop flavor can be achieved, but also some bittering."

For Pelican Pub & Brewery's Kiwanda Cream Ale, brewmaster Darron Welch adds the beer's only hop addition at flameout. Welch gets about 25 IBUs from adding roughly 0.75 pounds/bbl (0.34 kg/bbl) of Mount Hood hops at flameout, then allowing a 30-minute whirlpool stage. This means that Darron is getting roughly 16 percent utilization on his 15 bbl system for a 1.049 specific gravity wort. As mentioned, in a homebrewers hop stand, the 5-gallon (19-L) kettle is going to cool much faster and therefore create lower utilization rates. Brad Smith, creator of the BeerSmith brewing calculator, gives this advice to homebrewers: "Something in the 10 percent range is not a bad estimate if hops are added near boiling and left in during the cooldown period." In *BYO's* testing of extended hop stands in 11-gallon (42 L) batches, a 10 percent utilization rate for whirlpool hops seems reasonable.

GIVING IT A TRY

While hop-forward beers can benefit from this technique, any beer where some hop nose is desired is also a good candidate. For low-IBU beers, you can completely skip a separate boil hop addition, as discussed, or you can add a tiny bittering charge of hops to help break the surface tension of the beer and then add all or the majority of the IBU contribution at knockout, with the 10 to 15 percent utilization in mind.

The second factor to consider is the length of your hop stand. There are no right or wrong answers, but anywhere from 10 minutes to 90 minutes is reasonable. For most supercharged, hop-forward beers, hop stands will run 45 to 60 minutes. For a midrange hop profile, such as an American pale ale, you can usually shorten that stand to 30 minutes. If the beer is not to be hop forward, nor does it need significant IBUs from the hop stand, then a 10- to 15-minute hop stand usually will suffice.

TIPS FROM THE PROS

LAYERING AROMA HOP ADDITIONS AT PARISH BREWING CO.

By Ryan Speyrer

We focus on making flavorful base wort with generous late-hopping additions in the kettle. Many of our beers don't receive hops until flameout at the earliest, and these additions typically range from 1.25–2 lbs./bbl (0.65–1 oz./gal. or 4.8–8.5 g/L) depending on the recipe. We also perform "sub-iso" additions on some beers, including all of our hazy IPAs, where we cool the wort going into the whirlpool tank to below isomerization temperature and add hops at a similar rate. This technique seems to impart richer citrus and melon flavors from hops that already exhibit those potential flavors.

The active fermentation off-gassing by the yeast will have a scrubbing effect on some of those more volatile hop aroma compounds, so some loss will be observed between the brewhouse and post-fermentation. Conversely, yeast contact with the hops can provide interesting flavor changes via biotransformation whereby certain yeast strains can chemically alter hop aroma oils into new compounds with different flavor profiles.

Most of our hop load goes into the dry hop post-fermentation where we will use from 2 up to 10 lbs./bbl (1–5 oz./gal. or 8.5–42 g/L). Moving beyond this threshold we have found a significant rise in vegetal, grassy, and hop astringency imparted to the beer (not to mention tremendous yield losses). When primary fermentation is complete, we cold crash the tank, remove the yeast we plan to reuse, and dump the trub from the bottom of the tank before dry hopping. Keeping contact with the hops warm will impart undesirable grassy and vegetal notes to the beer, especially with very heavily dry-hopped beers, so we consider the cold-crashing step to be more important than simply looking at contact time. We've found 2 days is all that's needed for most of the flavor to come from dry hop additions.

The later you add the hops in the brewhouse, the more aromatics will be retained in the wort. There is no getting around dry hopping to get maximum hop aroma in your beer, however.

Note: *If you're looking to turn your hop stand into more of a whirlpool along the lines of a proper brewery, you have a couple options. If your brewing system has a pump, you may opt to set up a tangential inlet for your kettle to allow the pump to perform the whirlpool for you. Keep in mind that you do not need a vigorous whirlpool—just a simple spinning of the wort. If you do not have a pump, a simple spoon or paddle will work to achieve a similar result.*

Three temperature profiles that seem to be popular among homebrewers are just off boil range—190 to 212°F (88 to 100°C)—the sub-isomerization range—160 to 170°F (71 to 77°C)—and a tepid hop stand range—140 to 150°F (60 to 66°C). The 190 to 212°F (88 to 100°C) range will allow essential oils with higher flashpoints an easier time to solubulize into the wort and also will allow some alpha acid isomerization to occur with the best estimates of between 5 to 15 percent utilization. Some homebrewers will keep their kettle burner on low to keep the temperature of the wort elevated above 200°F (93°C) during their extended hop stands, which would better emulate the conditions in commercial whirlpools. A hop stand in the 160 to 170°F (71 to 77°C) range will basically shut down the alpha acid isomerization reaction, and the lower temperatures will reduce the vaporization of the essential oils. Homebrewers can use their wort chillers to bring the wort down to this range before adding the knockout hops or they can add a second dose of knockout hops. The 140 to 150°F (60 to 66°C) range will once again reduce vaporization of the low flashpoint oils, but it may take longer to get the same amount of essential oils extracted.

DRY HOP CONSIDERATIONS

Another factor to consider is how to handle dry hopping your hop-forward beers if you employ an extended hop stand. Rock Bottom Restaurant & Brewery performed an extensive study on hop stands and dry hopping under the guidance of the Portland, Oregon, brewmaster at the time, Van Havig (now of Gigantic Brewing Co. in Portland). The study was published by the Master Brewers Association of the Americas *Technical Quarterly* and considered beers that were hopped in four different ways: short hop stand (50 minutes) and no dry hops, long hop stand (80 minutes) and no dry hops, no hop stand and just dry hops, and finally half the hops in hop stand (80 minutes) and half the hops for dry hopping. Beers produced using exclusively hop stands and the beers produced using exclusively dry hops will both result in well-developed hop characteristics, but there were some nuances.

The long hop stand developed more hop flavor and aroma than the short hop stand, indicating that essential oils were still soaking into the wort after 50 minutes. The exclusively dry-hopped beer received its best marks in the aroma department, higher than the hop stand beers, but scored lower for its hop flavor. The beers where only half of the hops were added for the hop stand and half were added for aroma ended up scoring high in both departments. Havig's study also showed that adding 1 pound/bbl (0.45 kg/bbl) Amarillo dry hops produced the same amount of hop aroma as $^1/_2$ pound/bbl (0.23 kg/bbl), indicating diminishing returns at higher dry hop rates.

So if you are just giving this technique a try, here is a suggestion based on the study's findings. Take all the hops you plan to add for late-addition hops and dry hops and cut them in half. Add half at knockout and the second half as a dry hop addition. Again, don't feel the need to go overboard with these additions.

DRY HOPPING

What is it about a beer like Russian River Brewing Company's Pliny the Elder that is so striking? To anyone who has tasted or brewed a beer such as Pliny the Elder (see recipe on page 220), it's the hop oils that knock your olfactory senses sideways. Thanks in a large part to groundbreaking work by Thomas Shellhammer, Nor'Wester Endowed Professor of Fermentation Science at Oregon State University, and his former student Peter Wolfe, who now resides as brewing scientist at Anheuser-Busch InBev, our understanding of the extraction process has advanced that much more. Using some advanced dry hopping techniques, we can use this science at home to achieve hop bliss.

DRY HOP BASICS

The definition of dry hopping is adding hops to your wort after the boil, after it's been cooled—and typically when fermentation is almost finished or completely finished. This is different from other hop additions, even finish hopping or using a hopback, because the hops aren't exposed to heat when dry hopping and therefore do not cook.

Dry hopping is essential for modern IPAs.

Without heat, the hop alpha acids do not isomerize and hence do not add bitterness like hops added to hot wort, which contribute to bittering.

There are a few different ways to dry hop, depending on the variety and form of hops you plan to use. Selecting a hop variety for dry hopping is subjective. Some brewers prefer using aroma hops that have a low to medium alpha acid rating, such as Cascade, East Kent Golding, Saaz, Glacier, or Willamette. Other hop-happy brewers like the character of more aggressive high-alpha varieties, such as Chinook and Simcoe. There are no hard and fast rules.

Choose a form of hops that you will feel comfortable handling for your brewing setup. Whole leaf and pellet hops are both easy to add to a carboy, and if you're using a fermenter with a larger opening, you can put your hops in a sanitized grain bag and steep them like tea. A good rule of thumb is to use around 0.5 to 2 ounces of hops for a 5-gallon (19 L) batch, depending on the variety—though aggressive IPAs and double IPAs will often call for much more.

Once you've chosen your hops, you can add them to your cooled wort. Some brewers like to add the hops to the primary fermenter. If you choose to add them at this point, you need to account for some aroma loss from the release of CO_2 by adding more hops than you might think you need.

Most brewers add their hops to the secondary fermenter, or at least after primary fermentation has finished, which avoids aroma loss problems. Plus, beer in the secondary has finished fermenting and has a lower pH, so there is less of a risk of contamination from anything on the hops (which is a low risk either way).

Additional details follow, but in general you'll steep your hops for a week or two if you're using ale temperatures and 2 to 3 weeks at lager temperatures. Keep tasting your beer throughout the process until you find the flavor you want, if you're new to dry hopping. Then, rack your beer away from the hops or remove the hop bag and you're ready to start bottling or kegging.

Now, on to the more advanced stuff!

IDENTIFYING THE PLAYERS

Extracting hop oils and other aroma components from hop cones is the driving force behind dry hopping beers. The exact number of different hop oils found in the lupulin glands of hops has been found to be nearly five hundred unique forms. Those hundreds of hop oils can be split into three major classification groups: hydrocarbons, oxygenated hydrocarbons, and sulfur-containing compounds. The hydrocarbon group makes up well over half of the hop oils by weight in a hop cone, and most hop-heads might know them as terpenes. Hop oils such as myrecene, pinene, and humulene are just three examples of terpenes (hydrocarbons) that many folks versed in this subject would recognize. But don't let names fool you. For example, humulene has been shown to come in seventeen distinguishable forms in a hop cone, each slightly different than the other (welcome to the wonderful world of organic chemistry).

Oxygenated hydrocarbons include the terpenoids. Terpenes and terpenoids have a very similar skeleton structure, but the terpenoids will include an oxygen group. Esters and alcohol groups fall into this category with familiar hop oils such as linalool, geraniol, and citronellol. This group will come into play later when discussing glycosides, so pay mind to this group. Finally there are the sulfur compounds such as thiols like 4mmP, a polarizing compound some beer drinkers liken to cat pee while others perceive as tropical fruit aromas. Many recent studies have shown that sulfur compounds may play a bigger role in the hop characteristics than thought previously, given the very low sensory threshold of many of these compounds. Even though they make up less than 1 percent of the hops oils, the ultimate weight they carry into the beer may be rather hefty.

Just outside of the hop oil world, but very relevant to this discussion, you'll find the glycosides. Glycosides are in fact a combination of a terpenoid (see earlier) with a sugar molecule (glucose). Peter Wolfe explains how glycosides can play a prominent role in the aroma of beer: glycosides are tied together with a *relatively unstable* bond (an ester bond) between the glucose group and the terpenoid. In beer, this ester bond can hydrolyze (break apart) and release the terpenoid and the glucose to the solution. So if you can hydrolyze the glycoside, you increase the terpenoids in solution.

Note the emphasis on relatively unstable: the bond break won't happen by itself. It needs a push. That can happen in two ways. The first is a spontaneous reaction based on the pH of the solution. The lower the pH, the faster the spontaneous hydrolysis reaction can occur. This is convenient since beer pH is much lower than wort pH, and this spontaneous reaction won't occur until pH is down near 4.4. The pH of beer is generally 4 to 4.2. The lower the pH, the faster this hydrolysis reaction occurs. The second way has been shown to occur thanks to yeast.

Hydrocarbons are generally the most volatile of the hop oils while their oxygenated cousins are less so. Glycosides are not volatile at all. Many brewers will look to the hop oil's flash point to gauge the volatility of the hop oil. Generally, the higher the flash point the less volatile the hop oil. Hop oils can generally be diminished in your beer by three means. First, it can be diminished by heat: the volatile oils can be vaporized more easily the warmer the solution. This most often occurs while the wort is still boiling or just after the boil. Second, it can be diminished by scrubbing. This most often occurs in the fermenter. Brewers tend to refer to the scrubbing process either when volatile oils are pushed out of the fermenter with any CO_2 escape or when oils stick to yeast membrane and are effectively dragged out of solution as the yeast flocculate (settle). Finally, hop oils can degrade by age or by oxygen. This occurs most often in the racking or bottling process and subsequent aging of the beer.

Now to put these concepts into play. With help from seven brewers: Vinnie Cilurzo, owner/brewmaster at Russian River Brewing Co. in Santa Rosa, California; Matt Brynildson, brewmaster at Firestone Walker Brewing Co. in Paso Robles, California; Jamil Zainasheff, owner/brewer at Heretic Brewing Co. in Fairfield, California; Ashton Lewis, master brewer at Springfield Brewing Co. in Springfield, Missouri; Josh Pfriem, owner/brewmaster at Pfriem Family Brewers in Hood River, Oregon; John Kimmich, owner/head brewer at The Alchemist in Waterbury, Vermont; and Jack Hendler, co-owner/head brewer at Jack's Abby Brewing in Framingham, Massachusetts.

KNOW THINE ENEMY (O_2)

Every brewer and scientist interviewed for this section had the same piece of advice to homebrewers: focus on minimizing oxygen uptake post-fermentation. So let's talk about this first. John Kimmich has focused a lot of time and energy toward making sure that the dissolved oxygen (DO) of his beers is minimized. While it may not be the only reason his beer, Heady Topper, is one of the highest ranked beers in the world, it certainly helps. How little DO can be found in Heady Topper? One time a quality assurance employee from John's canning company came to test his beers. The tester soon ran back to his car as he thought his DO meter was broken. He had measured 1 part per billion (ppb) DO in the Alchemist's brite tank (a vessel that is loosely the commercial version of a homebrewer's bottling bucket). He had never seen numbers that low! While John's processes aren't all public knowledge, one thing we do know is that John is adamantly opposed to filtering his beers. And the yeast left in solution can act as a buffer against any oxygen uptake. This is one reason you may not want to filter your beers when brewing a hop-forward beer.

So how and when is oxygen going to be introduced into your beer after fermentation finishes? There are two principle ways so long as you are using proper brewing equipment and not aging for long periods of time. The first potential culprit for O_2 ingression is from racking. To help you solve the racking dilemma, homebrewers have four options: let's call them the bronze, silver, gold, and platinum option. The bronze option is for homebrewers who don't have CO_2 on hand. Rack as gently as possible and, if possible, rack before the yeast has finished primary fermentation. This will allow a new blanket of CO_2 to develop after racking. The silver option is for brewers with access to CO_2 who can purge their receiving vessel with a shot of CO_2.

One physical trait of CO_2 to keep in mind is that it is slightly heavier than air (N_2 and O_2) so it will settle on the bottom and help create a blanket as long as there isn't too much agitation. So you may not need to purge all oxygen, just enough to create a healthy blanket. Our own Ashton Lewis isn't as convinced, saying, "Small differences in temperature create convective currents that mix gases. It may be true that there may be a greater concentration of carbon dioxide at the bottom of the vessel, but there is still more than enough oxygen in most cases of simple blanketing to oxidize beer."

So finally we get to the gold and platinum options, which keep the system closed, and thus it's basically impossible for oxygen to get introduced to the beer. See the sidebar on page 96 for details on these closed-system racking techniques.

The second potential culprit for the introduction of O_2 is the addition of any post-fermentation ingredient such as dry hops, coffee, cocoa nibs, or anything else that brewers may add to their beer. Pro brewers have created some highly inventive devices to deliver dry hops to their fermenters in an oxygen-free system. For example, you may have heard of the hop cannon (which will shoot pellets into the fermenter via CO_2) or Sierra Nevada's hop torpedo (an inline recirculation system that passes the finished beer through the dry hops before returning the beer back to the fermenter). Peter Wolfe and Thomas Shellhammer both agreed that these devices were overkill on the homebrew scale as homebrewers can get away with eliminating most oxygen from the dry hop process with less effort. Thomas Shellhammer's advice, especially with whole cone hops, is "to ensure the hops are free of residual oxygen before adding to the beer. Do this by vacuum packing them first; better yet, vacuum pack and then gas flush them with nitrogen or carbon dioxide. Another approach is to submerge the hops in cold sterile water, then transfer the whole lot to the beer." Ashton Lewis adds that working with deaerated water is a huge benefit if you can get it. Boiling the water will help deaerate but won't remove all residual O_2.

Peter Wolfe's suggested approach is slightly different. "I like to rack my beer onto the dry hops when there are still a few gravity points left to go until terminal gravity is reached," he said. This works on the principle that active yeast will help absorb some of the oxygen uptake that has occurred during transfer. And finally for those homebrewers who try to emulate the recirculation methods utilized by probrewers, Vinnie Cilurzo adds that you have to make sure you are not introducing any oxygen into the system. If you can have your beer read 1 ppb DO after racking and adding dry hops, you've mastered this enemy.

CHOOSING YOUR WEAPON

The age-old discussion is: Which is better for dry hopping in your hop-forward beers—pellets or whole hops? However, the most important aspect is that you are choosing aromatic hops that are pleasing to your olfactory senses. To put it simply, "the hops need to smell awesome in order to make awesome beer," says Matt Brynildson. Vinnie Cilurzo's advice is to "watch out for hops that have onion/garlic character. You will never be able to get rid of this from the hop."

Still, some brewers prefer whole hops and some prefer pellets. Tests run by Peter Wolfe and Thomas Shellhammer point to pellets holding a slight edge over the unprocessed form. What they found is that pellets actually contain less essential oils than whole cone hops, most likely due to the pelletizing process. But oils from pellets are more easily extracted into beer during dry hopping, leading to slightly more aroma than when using whole cone hops. Also the oils in pellets were extracted faster than whole cones. Pellets have better storage life and are easy to work with for dry hopping since they submerse themselves, break apart, and fall to the bottom, so you can easily rack your beer off of them.

TIPS FROM *BYO*

CLOSED-SYSTEM RACKING

Closed-system racking is the Holy Grail for meticulous brewers. Peter Wolfe highly recommends homebrewers try to create a closed-racking system, especially for hop-forward beers. The two closed-system racking options are most easily done with either a conical fermenter or carboys with a racking cane fixed into a hood cap, but it could also be rigged up with modifications to a bucket lid. There are two objectives to overcome to close your system off. The first goal is to purge the receiving vessel with carbon dioxide (or other inert gas). The second is to make sure you are adding CO_2 into the top of the sending vessel.

The method is what we will call the gold method. First you slowly purge the receiving vessel (such as a Corny keg) with CO_2 for about 1 minute on very low pressure (1 to 2 psi). Once the receiving vessel is purged, now it is time to rack. Using the carboy hood cap with a racking cane, push CO_2 into the sending carboy, forcing the beer out of the carboy and into the purged receiving vessel (**see Figure 1**).

Now for the platinum method. The first step in this method is to completely fill the receiving vessel with liquid such as a dilute iodophor or Star

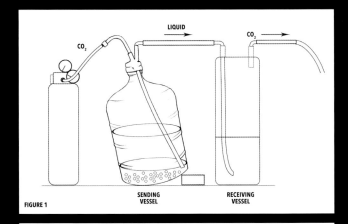

FIGURE 1

SENDING VESSEL / RECEIVING VESSEL

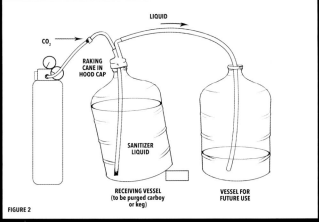

FIGURE 2

RECEIVING VESSEL (to be purged carboy or keg) / VESSEL FOR FUTURE USE

If you are going to be using whole cone hops, you need to deal with two potential factors that could hurt the beer. First is to make sure that if O_2 is found in the cones, that you flush them with CO_2, nitrogen (N_2), or sterile water before adding the beer (see Thomas Shellhammer's earlier advice). Racking the beer on top of the flushed hops is then the preferred method once you have them flushed. The second factor is that whole cones float and you need a way to submerse them without adding oxygen. A sanitized bag

San solution. The second step is to push all the liquid out of the vessel with CO_2 (**see Figure 2**). *BYO's* Ashton Lewis adds, "This can be done in a carboy without adding any real pressure if the water is siphoned out of the carboy and displaced by very low pressure gas. We use this method at Springfield Brewing Company because purging was not working for us and we switched to water flooding about 10 years ago. It also uses less gas, but adds time to the schedule." This method works incredibly comfortably and efficiently when performed on the *BYO* brewing system. We push the dilute sanitizing solution from one keg or carboy to another using CO_2. Now we have a receiving vessel with 100 percent CO_2.

Using a fermenter with a spigot at the bottom allows gravity to feed the beer from the fermenter to the keg or carboy, displacing the CO_2 back from the receiving vessel back into the fermenter, making sure that no oxygen can enter the system (**see Figure 3**). If for some reason you don't have the luxury of gravity, pushing via your CO_2 regulator works just as well. Just be sure to start very slow and low with CO_2 if doing this, as you can overpressurize the system very easily. Carboy caps and bucket lids were not meant to withstand pressure. Start with your regulator turned all the way to zero and very slowly dial up the pressure. Two psi is more than enough pressure to apply.

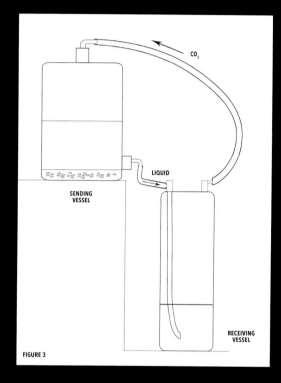

FIGURE 3

that is weighted down with sanitized stainless washers or marbles can be used to make sure that the hops remain submerged. Whole cone hops could also be added before primary fermentation is finished and then gently pushed down with a sanitized paddle to completely submerse them.

Despite what research shows, it is all about what works best for your system. Some homebrewers prefer to work with whole hops while others prefer to work with pellets. Both can produce award-winning beers.

TIMING IS EVERYTHING

Many brewers try to eliminate as much yeast as possible from suspension before adding dry hops. Vinnie Cilurzo introduced that concept to many homebrewers in a sound bite that has been oft repeated. The reason was simple, as stated earlier in this article: yeast can strip hop oils from solution. Thus, many homebrewers either rack their beer to secondary before adding dry hops or add dry hops only after adding a fining agent such as Polyclar or Biofine. Yet other homebrewers advocate adding dry hops while yeast is still active. The term *biotransformation of hop oils* is something you may have heard to back up this line of reasoning.

Two of the seven brewers interviewed for this section add their dry hops while yeast was still active. On the one side of the coin, Josh Pfriem states that "the constant nucleation from fermentation scrubs away some of the hop aromas that you are trying to achieve." Matt Brynildson, on the other hand, adds his hops during fermentation, and the hardware garnered for his hop-forward beers should make any brewer rethink their approach. Matt has three reasons for his method: "This is to take advantage of the active yeast for (1) dissolved oxygen protection; (2) natural mixing, which we believe helps in better extraction of wanted oils; and (3) biotransformation of hop oil compounds." But he also warns of the potential pitfall of adding dry hops to an active fermentation. "The dreaded 'beer volcano' can happen easily resulting in beer loss," he says.

So what are biotransformations anyway? Once again we turn to Peter Wolfe to help dissect this term. He explains that when we talk about biotransformations on hop compounds we are talking about oil components that yeast have modified. An important aspect is that we are talking mainly about terpenoids and glycosides. Terpenes are rarely affected by biotransformations. Biotransformations of hop compounds in beer can occur in two forms. The first is fairly straightforward when one compound is transformed into another. An example of this would be the transformation of geraniol to ß-citronellol. The second biotransformation is the hydrolysis of the glycosides, which was introduced earlier. Certain yeast strains have shown the ability to transform nonaromatic glycosides into aromatic terpenoids. Shellhammer and Wolfe found that certain aromatic terpenoids increased their concentration over time in the presence of yeast. This may be one reason many people find bottle-conditioned or unfiltered beer to be superior to filtered beer.

CREATE THE PERFECT ENVIRONMENT

That brings us to the next big questions of dry hopping: finding the perfect time, temperature, and amount of hops. Many homebrewers were going as short as 3 days. Of all the brewers, only Vinnie Cilurzo went more than 5 days with his dry hops. Research by Wolfe and Shellhammer confirm that hop extraction occurs rapidly. In fact, in a recirculating system they found that most aroma compounds are extracted from pellet hops in a matter of hours. For pelletized dry hops added without recirculation, they found that full extraction occurred in 1 to 2 days while whole hops took closer to a week for full extraction.

The next variable to look at is temperature. Considering the rise of the IPL (India pale lager), the temperature of the beer and its effect on the oil extraction is something to keep in mind. So my first turn in this department was to Jack's Abby Brewing Brewmaster Jack Hendler, whose lagers have made waves in the US craft beer scene. Surprisingly, the average time Jack dry hops is only 3 to 4 days. He does increase his temperature to 55°F (13°C) for adding hops, but adds, "The cooler the temperature, the less aroma you'll pull from the hops." If you do plan to try an IPL of your

BREWING HAZY IPAS

New England IPAs have a blonde to golden color and are low in perceived bitterness, with lots of hop aroma (especially citrus, ripe tropical fruit, and stone fruit notes). They often have a creamy/silky palate, expressive aromas from yeast esters, and sufficiently high carbonation to fit with the body and to puff up into a white and wispy foam.

A good all-around New England IPA can be brewed incorporating the following pointers:

- Low carbonate water with a 2:1 ratio of chloride-to-sulfate.
- Keep the grist bill simple. Eighty percent pale malt, 10 percent flaked oats, and 10 percent flaked wheat works well. Depending on your mashing method, rice hulls can help with wort collection.
- Don't add too much hop bitterness to your wort during boiling; a modest addition at the beginning of the boil lays down a bit of bitterness on the blank canvas and helps suppress foaming during the boil. After the boil, knock the wort temperature down to about 176°F (80°C) before adding about a third of your aroma hops to hot wort. The reduced temperature will help keep isomerization to a minimum and allows for big additions for aroma.
- Use a yeast strain that is known for the style. There are a handful out there that work well, such as Wyeast's London Ale III and SafAle's S-04 or Imperial's Juice just to name a few. One of the keys to this style is hop biotransformation and this is largely a function of yeast strain.
- A good dry hopping schedule is to add a third of your aroma hops about 24 hours after vigorous fermentation begins, and the last third 48 hours later.

- If you have hops that seem off in terms of aroma quality or if they have aromas that you don't like, don't use them.
- Allow your beer to settle well before packaging into bottle or keg, and, like most other beer styles, do be concerned about oxidation.

Follow this basic guide and you will end up with a cloudy brew for sure. In the world of commercial brewing, consistency is very important, especially for packaging breweries, so haze stability is a thing. And it's a perplexing conundrum because cloudy beer is inherently unstable—haze particles tend to be denser than beer and gravity never takes a day off. The haze in these beers is not from yeast or "chunky stuff" floating about in the beer. The cloudy appearance of this style is the product of protein and polyphenol (tannin) interactions, and it seems that a significant portion of these haze particles are not very dense and tend to stay bobbing about in their hoppy homes. Although most "hazies" contain somewhere between 10-30 percent flaked grain adjuncts that do bring protein to the party, the main haze ingredient is the big boost of polyphenols that comes with the massive hop additions.

Speaking of massive hop additions, these are not inexpensive beers to brew. Jean-Claude Tetreault, a leader in the hazy IPA style at Trillium Brewing Company, has used more than 10 lbs./bbl (5 oz. per gallon/37 grams per L) in some of his dry hop additions. On a homebrew scale, these beers generally require a minimum dry hop addition of 0.5 oz./gallon (3.7g/L) and go up dramatically from there.

own, Jack has some further dry hopping suggestions. "You'll need to reevaluate dry hop addition quantities, because the dry hop aroma will be highlighted more than in an ale. You may find a different or smaller quantity gets the aroma you're looking for," he said.

So how much dry hops should you add? Obviously that is completely dependent on what you are trying to achieve with your beer. But keep in mind that sometimes, more isn't always better. In terms of dry hopping, the more you add, the less net gain you add with each additional increment. In fact you may find that you are detracting from a certain nuanced characteristic of the beer if you overwhelm it with another characteristic. Finding the right balance of hop oils of a varietal or a blend is key. If you've had Heady Topper before, you may be surprised to learn that John Kimmich dry hops with under 4 ounces (113 g) per 5 gallons (19 L).

IS LAYERING THE KEY?

Only two of the seven pro brewers didn't add their dry hops in stages. One that was surprising was John Kimmich, who adds all his dry hops in one big charge for his Imperial IPA. Jamil Zainasheff pointed out an important nuance: "The main reason is that we're dry hopping into cylindroconical fermenters. The bottom is a narrow cone, which means that when the hops drop to the bottom, it results in a smaller surface area." He said not to worry about layering in dry hops on a small scale. Peter Wolfe also weighed in on the topic, saying especially if a homebrewer is using a flat-bottomed fermenter, there is little reason to layer in your hops; the surface-area-to-volume ratio is much greater on a homebrew scale.

There is one reason that homebrewers may want to layer in their dry hops. As stated earlier, Matt Brynildson adds dry hops while active fermentation is still ongoing. For some of his beers, he will add a first dry hop charge near the termination of active fermentation and a second addition after flocculation has occurred. He backs up his approach with the concept to "take advantage of both conditions (1) with yeast and (2) without yeast influence." So if you plan to add your first dry hop charge near the end of active fermentation, you may also want to take advantage of this two-layered dry hop approach. Otherwise, if you are planning on waiting until most yeast has settled from the beer, then one stage dry hops are all you need.

YEAST

Yeast is one of the four essential ingredients in beer, and its job is to transform wort into beer. Along the way, depending on the choice of strain, it will add varying amounts of flavors and aroma—as many as six hundred different compounds—to the beer. Yeast have three phases in their life cycle: adaptation, high growth, and stationary. Adaptation phase is where they take in oxygen and build sterols and other lipids, assess the sugar composition, and build enzymes, etc. Once those activities are done, they start the high-growth phase: eating and reproducing. The number of cell divisions is limited by their lipid reserves they made during adaptation. These reserves are shared with each daughter cell. When those lipid reserves are exhausted, the cell stops reproducing. In addition, when those reserves are exhausted, the cell is old and cannot eat or excrete waste efficiently across its cell membrane. A yeast cell typically can reproduce about four times during a typical fermentation; after that it is old and tired and tends to enter stationary phase where it shuts down most of its metabolism and flocculates, waiting for the next batch of aerated wort. Stationary phase is essentially an inactivity phase, resting on the bottom.

CHOOSING YEAST STRAINS

We spend a lot of time classifying beer into certain styles, such as American pale ale and European dark lager. The yeast a brewer selects plays a big part in defining the flavor and aroma of most beer styles. In this section, we'll take a look at different types of yeast and how they influence the beers we brew with them.

ALE YEAST

Ale has been brewed since at least ancient Egyptian times. Ale yeast goes by the Latin name *Saccharomyces cerevisiae*, and this species includes bread yeast, distillers yeast, and many laboratory yeast strains. Ale yeasts are distinguished by their unique flavor production. The use of bread yeast or other wild yeasts by brewers would result in phenolic-tasting beer. Ale yeast, as well as lager yeast, do not produce phenolic-tasting beer because

Whether you're buying liquid or dry yeast, always make sure your yeast is fresh.

they have a natural mutation that prevents them from producing phenolic off-flavors. (Specifically, brewer's yeasts—sometimes called POF (-) strains—lack ferulic acid decarboxylase, the enzyme that decarboxylates ferulic acid to produce four-vinyl guaiacol.) Ale yeasts do what a brewer wants: they ferment quickly, consume the correct profile of sugars, tolerate moderate alcohol levels, and survive the anaerobic conditions of fermentation.

There are a huge variety of ale yeast strains. In fact, all wheat and Belgian strains are classified as ale yeast. For the purposes of this section, however, they will be treated separately. Because of the large variety of ale yeast strains, there are many differences in performance among these yeasts. They flocculate differently, attenuate differently, and produce different flavor profiles.

They do have some similarities, however. Almost all ale strains have an ideal fermentation temperature that hovers around 68°F (20°C). Most ale yeast will tolerate conditions to 95°F (35°C), but they produce the best flavors when they ferment at or near 68°F (20°C). Flavors that ale yeasts produce are varied. If they produce a small quantity of these flavor compounds, they are known as clean fermenters. The more esters and fusel alcohols, the fruitier the yeast is considered. The late George Fix—in his book, *An Analysis of Brewing Techniques*—uniquely labeled these as group 1 and group 2 ale yeasts, respectively. Examples of clean group 1 yeasts include White Labs WLP001 (California Ale) and Wyeast 1056 (American Ale), White Labs WLP029 (German Ale/Kölsch), Wyeast 1007 (German Ale), and 2565 (Kölsch), as well as WLP051 (California V Ale) and Wyeast

1272 (American Ale II). The clean fermenting ale strains are very popular, because they can produce lager-like beers using ale techniques and fermentation times. They usually ferment a little slower than other ale yeast and exhibit medium flocculation properties, which ensure they will be in the beer long enough to condition it properly. They can also produce trace sulfur, but not as much as lager yeast strains.

Ale yeasts are famous for their ability to top ferment. After the first 12 hours of fermentation, many ale yeast strains will rise to the surface and ferment from the top of the beer for 3 to 4 days. This allows brewers to collect the yeast from the top, a practice called top cropping. The advantage of top cropping is a great crop of yeast, healthy and with little protein mixed in. The disadvantage of this method is the exposure to the environment. If the fermentation room is not sanitary, the yeast can easily be contaminated. Few homebrewers top crop, but I suggest giving it a try. Homebrewers who ferment in glass carboys will have trouble cropping the yeast through the small opening, but if fermenting in plastic buckets, this can be done with good cleaning practices. It is worth doing as an experiment, and you the brewer can evaluate its effectiveness.

Ale yeast that produce fruitier beers are less versatile but extremely interesting. The strains WLP002 (English Ale), WLP004 (Irish Ale), and WLP005 (British Ale) are examples of fruity group 2 yeasts—Wyeast strains include 1028 (London Ale), 1968 (London ESB Ale), 1084 (Irish Ale), and 1187 (Ringwood Ale). These are yeasts of distinction and can add a lot of character to your beer. They ferment at the same temperature as clean fermenters, but in doing so create more compounds that are excreted from the yeast cell. They usually flocculate quickly, which aids in leaving acetaldehydes and diacetyl in solution. It's fun to produce a beer with highly flocculent yeast because it looks different while fermenting. (Lots of chunks!) Then the yeast drops out right when fermentation is done. The beer can be bottled and consumed rapidly. These strains usually do not top crop as well because they flocculate too quickly.

For ale yeasts, a pitching rate of 5 to 10 million cells per milliliter promotes cell growth and good beer flavor. For a 5-gallon (19 L) batch of beer, this corresponds to the amount of yeast from a 1 to 2 quart (about 1 to 2 L) yeast starter.

LAGER YEAST

Lager yeasts ferment best at colder temperatures than ale yeasts—in the 50 to 55°F (10 to 13°C) range. Lager yeast is currently classified as *Saccharomyces pastorianus*. This special yeast was first isolated in the Carlsberg Laboratories under the direction of Emil Christian Hansen, in 1881. Hansen was the first to develop pure culture techniques, techniques that we still use today in microbiology laboratories. Not only was Hansen able to grow this new yeast, lager yeast, in pure form, but he was able to store it for long periods of time on a combination of wort and agar, which creates a semihard surface. This long-term storage allowed lager yeast to be transported all over the world, and soon lager brewing overtook ale brewing worldwide.

Why did lager beer become so popular? At the time lager yeast was discovered, most ale fermentations contained some wild yeast and bacteria and the resulting beer had a very short shelf life. Lager beer could be fermented cool, which suppressed the growth of wild yeast and bacteria. (Modern lager brewers tend to have problems with *Pediococcus* because of the slower fermentation, but they likely have fewer problems than pre-modern ale brewers.) Lager beer therefore had a longer shelf life, which meant greater distribution area and increased sales. Breweries probably began to switch to lager brewing to increase their sales.

But what makes lager yeast so different from ale yeast? Unlike many ale yeasts, lager yeast does not usually collect on the top of the fermenting beer. Lager yeasts are known as bottom fermenters because of their nature to ferment from the bottom of the tank. But as in everything else in science, there are always exceptions! For example, White Labs WLP800 (Pilsner Lager) forms a yeast cake on the top of the fermentation even though it is a true lager yeast. While most lager yeast ferment from the bottom, they are not known as high flocculators. In fact, most of the yeast stays in suspension and most lager strains are low to medium flocculators. Lager yeast needs to stay in suspension in order to lager the beer, aging it with some yeast in order to reduce the sulfur and diacetyl levels produced during the cold fermentation.

The cold fermentation has many consequences for lager beer. Since the yeasts ferment in a cool environment, usually 50 to 55°F (10 to 13°C), they produce fewer esters and fusel alcohols. But the cool temperature keeps more sulfur in solution and makes it harder for the yeast to absorb diacetyl. A good diacetyl rest near the end of fermentation will greatly reduce this. Here is a good procedure for a diacetyl rest: ferment at 50 to 55°F (10 to 13°C). When the beer reaches a specific gravity of 1.022 to 1.020, let the fermentation temperature rise to a maximum of 68°F (20°C). The fermentation will reach completion, usually in the specific gravity range of 1.010 to 1.014. Let it sit at this temperature for 3 to 5 days post terminal gravity, then cool over the next day to 50°F (10°C). Keep at 50°F (10°C) for 1 day, then lower to lagering temperature, usually 41 to 45°F (5.0 to 7.2°C).

The optimal pitching rate for lagers is roughly double the rate for ales, between 15–20 million cells per milliliter.

WHEAT BEER

Traditional European wheat beers use special yeast strains, which produce lots of flavor. The flavors are what we usually classify with wild yeast, phenolic and clove. But these strains produce a pleasing amount of flavors that blend well with the other wheat beer ingredients. There are not a lot of different wheat beer strains, perhaps half a dozen. They differ in small ways due to different flavor compounds. For example, White Labs has two regular wheat beer strain offerings. One, WLP300 (Hefeweizen), has a dominant banana ester character that can be dialed in and out with fermentation temperature. The other, WLP380 (Hefeweizen IV), produces very little banana ester regardless of fermentation temperature. For WLP300, temperatures in the 65 to 68°F (18 to 20°C) range yield little banana; fermenting above 70°F (21°C) greatly increases the level of banana esters. Wyeast also carries wheat beer strains, including 3068 (Weihenstephan Wheat), 3333 (German Wheat), and 3638 (Bavarian Wheat). Wheat beer strains produce little alpha acetolactate, the precursor to diacetyl, and they absorb diacetyl quickly. So butterscotch flavors are rarely a problem. Wheat strains will also produce sulfur. It is important to let the fermentation go to completion before capping it. Some brewers like to cap the fermentation near the end to trap the remaining CO_2 as a way to carbonate the beer. If you do this, you will trap the sulfur in the beer and it will never go away. This applies to lager brewing as well. It takes approximately 24 hours post-fermentation at fermentation temperature to scrub all the sulfur out of solution.

Most wheat beer strains do not flocculate well. This is a desired characteristic, which leaves the traditional cloudiness of wheat beers. Wheat malt, with higher protein content, also contributes to this. You still want the yeast to drop out some; otherwise the beer will be milky like a yeast culture. The pitching rate for wheat yeasts is the same as that for ale yeasts.

BELGIAN STRAINS

Many Belgian beers are brewed with unique yeast strains. Many different strains are used, so it is impossible to generalize about them. You could make a Belgian-style wit beer with a normal lager yeast if you like, but it will just not have the nose and taste of a Belgian wit fermented with an authentic Belgian wit yeast strain. These strains usually produce a lot of phenol and clove flavors, as with wheat beer yeast. Many Belgian-style yeast go beyond just phenol and clove and produce a lot of esters, fusel alcohols, and earthy flavors. The balance of these compounds helps determine the flavor profile of these strains. These strains would be ones that are used to make Belgian-style Trappist beers, for example. Many Belgian strains, as with wheat beer strains, do not flocculate well. I recommend a higher pitching rate—between 10 to 15 million cells per milliliter—for most Belgian strains.

BRETTANOMYCES

Brettanomyces, often referred to simply as *Brett*, is a key component in brewing sour or "wild" beers. It is a genus of yeast, just like *Saccharomyces cerevisiae*, and just like *Sacch*, *Brett* encompasses many different species that can produce a wide spectrum of flavors in beer.

While *Brett* is a star in modern experimental craft brewing, its history comes from pre- and early twentieth-century beers that were fermented and aged in wooden tanks and barrels. It was classified at the Carlsberg brewery in Copenhagen, Denmark, in 1904, and is a classic component in Old World styles such as lambic, gueuze, saison, and Flanders red. Many modern craft brewers now use *Brett* in combination with *Saccharomyces cerevisiae* and lactic acid producing bacteria (*Lactobacillus*, *Pediococcus*), such as the wild ales from Russian River Brewing Co. (see page 220), New Belgium Brewing Co. (see page 211), or Jolly Pumpkin Artisan Ales (see page 202), while others brew with 100 percent *Brett*, such as Anchorage Brewing Company.

There are many strains of *Brett*—even more than *Saccharomyces*—but only a few are commercially available, and the two that are most commonly used for brewing beer are *B. anomalus* and *B. bruxellensis*. *Brett* differs from regular brewer's yeast because it has the ability to ferment dextrins (long chains of sugar molecules) that *Saccharomyces* can't break down; when this happens, it produces esters and phenols that give the beer characteristic *Brett* aromas and flavors. Some are welcome additions, such as hay, cherry, apple, and pineapple; others such as barnyard and horse blanket are desirable in some styles, but too much can be considered a fault. Some *Brett* effects are downright bad for your beer and will ruin a batch, such as medicinal aromas and acrid smoke.

Brett prefers a low pH environment, grows slowly, and can take months for it to ferment your beer and create *Brett*-specific characteristics. In his book *American Sour Beers*, Michael Tonsmeire emphasizes always pitching healthy *Brett* cells, especially when pitching them into a beer that has already been fermented with *Saccharomyces cerevisiae*. The proper pitching rate for *Brettanomyces* when pitching at the same time as *Saccharomyces*, or after a *Sacch* fermentation, ranges widely from 100 cells/mL to 2 million cells/mL. A 100 percent *Brett* fermentation requires the same pitching rate as a lager, or about twice the rate you would use for a regular ale fermentation. You can create a *Brett* yeast starter similarly as you would when making a starter with *Saccharomyces cerevisiae*, although give it an extra day or two to grow up (see instructions for making a yeast starter on page 42).

KVEIK YEAST AT BURIED ACORN BREWING CO. & EBB AND FLOW FERMENTATIONS

By Tim Shore and DeWayne Schaaf

Kveik provides a fast and clean fermentation, but the most attractive attribute about kveik strains are their ability to create ester combinations not attainable through other commercially available yeasts. Buried Acorn uses kveik for almost all their beers, and Ebb and Flow has also brewed all styles except those that require POF+ yeast, like saison. However, Kveik works well co-pitched with saison or Lithuanian yeasts that are POF+ and both love warmer temperatures. *Brett* and bacteria both play well with kveik for mixed fermentation styles. Honestly, there isn't much that they aren't good at if you let the concept of "proper" go.

Ebb and Flow is even brewing some kveik-fermented "faux lagers." Original cultures like Ebbegarden and Skare, or commercial cultures like Oslo from Bootleg Biology, ferment very clean and have marked lager-like profiles. DeWayne has played around with a number of fermentation temperatures, but has mostly kept it in the high-90s°F (mid-30s°C) to finish quickly before cold conditioning. The cold conditioning period greatly diminishes the exaggerated fruity ester profiles that kveik is known for.

Flavor profiles are strain-dependent, and it's difficult to modify expression of a single strain with traditional cellaring techniques. In Buried Acorn's experience, the profiles, although unique, are nearly identical fermented at 59°F or 108°F (15°C or 42°C) regardless of pitching rate. So, to get different profiles in different brands,

Tim relies on using different strains. There are a handful of commercial yeast labs that have some wonderful isolated kveik strains that can be swapped in and out depending on the beer.

Tim's favorite is an original mixed culture from Voss that he has kept alive and working for years. Voss creates a Grand Marnier kind of candied orange peel flavor that blends well with *Brett* barrels that are pineapple-forward. He uses Voss in an open fermenter at 104°F (40°C) and lets it ride—it's usually done fermenting and ready to transfer to barrels by the morning. This quick reduction of simple sugars keeps *Lactobacillus* from souring and leaves any acid to be produced by *Pediococcus*. Like most all yeast, kveik is not expressive in an already acidic environment.

DeWayne recommends resisting the temptation of brewing an IPA with your first kveik fermentation. Instead, start with an English mild or bitter, or perhaps a German Kölsch. Brew it in the summer so you do not have to worry about your chiller not getting your wort down to 62°F (17°C) or your closet not holding at 65°F (18°C). Don't be afraid to pitch it at 90–100°F (32–38°C) and keep it at that temperature until it is ready. These styles are better at helping you learn about the ester profiles that each kveik has to offer. A simple grist bill of Pilsner and pale malt (50/50) with 20 IBUs at 60 minutes will teach you a lot more about the capabilities of these yeasts than loading it up with hops. Save that for round two!

Brett cultures, both isolated as well as mixed with lactic acid producing bacteria, are readily available from commercial yeast companies, such as White Labs, Wyeast, East Coast Yeast, and others. Like *Saccharomyces cerevisiae*, *Brett* strains can produce all kinds of different outcomes depending on the strain and the conditions in the wort (presence of *Saccharomyces cerevisiae*, fermentation temperature, fermenting or aging in oak, and so on)—research, experiment, and have fun creating!

For more on homebrewing with *Brettanomyces*, check out Michael Tonsmeire's book *American Sour Beers*.

YEAST PITCHING RATES

A brewer's job is similar to that of an elementary school custodian: keep the place clean and make sure the environment is conducive for the little buggers to do what they need to do . . . which in this analogy is converting malt sugars into alcohol and CO_2 gas. We can ensure post-boil surfaces are free of contaminants; provide a quality, nutrient-rich wort; and foster favorable conditions (dissolved oxygen, fermentation temperature control, etc.), but the biggest determining factor that tips the balance between gallons or liters of awesome homebrew or pint after pint of blah homebrew is the health and happiness of the yeast doing the work of fermentation.

You have many factors to manage in conducting a good, healthy fermentation, but the first and most fundamental factor is inoculating the wort with a population of healthy yeast cells at the right dosage or pitching rate.

FINDING A BASELINE

Pitching rate is basically the amount of yeast one uses to inoculate cooled wort (pitching, in brewers' parlance), expressed as a ratio of the number of yeast cells to wort volume. A good baseline pitching rate is 6 million cells of healthy yeast per milliliter of wort. This rate is recommended for ales of average strength, which is what homebrewers make most of the time.

Most homebrewers make 5-gallon (19 L) batches. There are 18,927 (give or take) mL in 5 gallons (19 L), which works out to a target pitching rate of 113.5 billion cells for 5 gallons (19 L) of standard-gravity (<1.060 SG) wort that will be inoculated and fermented at ale temperatures (around 65°F/18°C or so). It's important to note that this is a guideline rather than a rule—there is room for variation and adjusting this number up or down depending on the style you're brewing and the temperature you use to ferment the beer (more on that later on in this section).

WHAT'S GOING ON IN THERE?

Before we move on to other considerations and theory into practice, let's take a step back . . . or rather a step way, way closer, and get a layman's overview of what happens when yeast meets wort, to help us understand why pitching rate is so important to the finished product.

A very simplistic perspective is that our brewer's yeast produces three things while colonizing its new home in our carboys and buckets: it makes more yeast cells, catabolic waste, and various metabolic by-products, such as malt sugars and dissolved O_2, through its use of wort resources.

More yeast is the straightforward result of cell reproduction, and that catabolic waste is CO_2 and alcohol (one fungus's trash is another man's treasure). The many various by-products yeast cells create during fermentation, then, include things such as esters, aldehydes, and phenols; the production (or lack thereof) of these by-products, and the level at which they pervade the batch, has a direct impact on the flavor and aroma of our finished beer.

QUALITY CONTROL AT FIRESTONE WALKER

By Jim Crooks

Many of the routine tests we do in the Firestone Walker lab can be done at the homebrew scale. The merits of a quality brew start with healthy yeast, and without a microscope and a hemacytometer to look at viability and density, you are just guessing about the quality of your yeast. This tool, as well as other tools like refractometers/hydrometers for checking gravity, thermometers, pH strips, beer color wheels, and basic water-testing kits, can be purchased inexpensively ($30–$100).

For the advanced homebrewer, setting up a field micro lab can be a little more involved but not out of the question. A homemade laminar hood and sterile sampling technique is the basic backbone of a small-scale micro program. WhiteLabs.com is a great homebrewing resource that offers a premade media test kit bundle that can be geared toward either anaerobic or aerobic bacteria as well as wild yeast. The White Labs site also offers a number of educational links as well as actual pictures of what bacteria look like when growing on the media. The potential cost savings to be had by doing a little micro investigation of your process can be well worth the cost of the tools versus the potential cost of losing a few batches to poor sanitation technique.

Some of the common brewing flaws we catch in the lab are ABV and color deviations from batch to batch. Although slight color deviation is not a pressing quality flaw, most people drink with their eyes, and color is one of the first noticeable attributes a customer will see, so we take it seriously. Rushing a sample through the program can also give rise to careless mistakes and often wastes more time by raising questions due to erroneous data. When in doubt, repeat your analysis.

A refractometer

WHAT DOES A PROPER PITCHING RATE TASTE LIKE?

Because those flavor-contributing metabolic by-products are created during the period of cell growth prior to what we lay homebrewers might think of as fermentation (kräusen, bubbling airlock, maybe some blowoff), the extent of the growth required of our yeast population determines how much of these compounds are present in the beer.

In a nutshell, higher pitching rate equals more yeast in, equals shorter growth phase, which equals lower esters, etc., and thus a cleaner, more neutral profile with less yeast character. Lower pitching rate equals less yeast in, which equals a longer growth phase and higher levels of esters, etc. and fruitier, funkier, and generally more yeast character in the beer's profile.

Besides ensuring reproducibility and consistency from batch to batch, controlling the pitch rate means we can also fine-tune the sensory profile of our pint. A clean, crisp Pils, where high levels of esters are a stylistic defect, mandates a high pitching rate—likewise for a Belgian tripel, where the combination of high-gravity wort with highly expressive yeast strains could get out of control quickly with fusel alcohols and high concentrations of esters and phenols. But maybe you're brewing a session bitter, or a weissbier with a hankering for a strong isoamyl acetate banana nose—a pitching rate on the lower end of the spectrum (without actually going out of the spectrum and underpitching) will help the yeast strain stand forward and put an authentic stamp on styles like these.

THE GOLDEN MEAN

Too much of anything is too much, as the man said—but not enough of anything isn't much good either. Underpitching a batch of homebrew, even for styles that benefit from some yeast character, can lead to higher-than-desirable levels of esters, fusel alcohols, or sulfur compounds, as well as more serious problems such as excessive amounts of diacetyl and higher-than-planned final gravity. Underpitching is a very common flaw in homebrewed beers. In the September 2010 issue of *Brew Your Own*, Brooklyn Brewery brewmaster Garrett Oliver, who has judged a great many homebrew competitions, said, "The most important factor to brewing any style, at home or professionally, is making sure you start from the very beginning with a healthy population of yeast. You should see the fermentation take off sooner rather than later. For example, for most ales, you ought to see a very active fermentation in under twelve hours. A beer from a struggling fermentation has a certain flavor. It's one of the main things that tends to distinguish what a professional might say is a homebrew flavor."

Overpitching, which might not seem like such a big deal on paper, can in practice significantly reduce ester production, cause overshooting of target final gravity, and speed fermentation up to create its own set of problematic flavors from yeast autolysis and can throw the beer out of style.

So to repeat: a good baseline pitching rate for standard-gravity (less than 1.060) wort that will be inoculated and fermented at ale temperatures (around 65°F/18°C) is 6 million cells of fresh yeast per milliliter.

HIGH-GRAVITY AND LAGER FERMENTATIONS

But what if you're going to be brewing north of 1.060 SG? Or if you're going to be fermenting south of 60°F (16°C)? Then you'd better bring more yeast cells. In addition to the considerations of flavor and aroma impact, high-gravity brewing creates added environmental stress for the yeast; the greater density of the wort means increased osmotic pressure on the cell walls, and the higher concentrations of alcohol as fermentation progresses becomes increasingly toxic to the yeast. For this reason, successful fermentation of a strong beer calls for starting out with more yeast than a standard-gravity brew.

FAST LAGERS

For decades, a common belief among homebrewers has been that it takes months to brew a lager. However, it's possible to complete a lager in as little as 2 weeks that is indistinguishable from a traditionally lagered beer.

There are different specific instructions and methods to creating a "fast lager," but they all follow the same basic concept of starting fermentation cool, as would be done traditionally, and then ramping up the temperature once fermentation has reached the first half (or so) of the expected total attenuation and the yeast has completed the growth phase (at this point the pathway to produce esters and fusels have become largely inactive) to speed fermentation without experiencing detrimental effects on the beer. (Refer back to page 100 for more on the three phases of yeast's life cycle.) By starting to raise the temperature when the fermentation is only partially done, you'll help the yeast be sure that the sugar is consumed sooner and that there will be plenty of lipids left to "nourish" the yeast and clean up by-products. Since yeast turns to by-products once the fermentable sugars are gone, you want those sugars fermented out before the yeast runs out of oomph.

One of the big proponents of the fast lager technique in the homebrew world was well-known homebrewer Mike "Tasty" McDole. Mike's process can be summarized like this:

1. Prior to fermentation, calculate your total expected gravity drop based on original gravity and yeast strain choice (75 to 80 percent is usual).

2. Chill the wort to 55°F (13°C) and pitch the yeast.

3. When the ferment is 50 percent complete, raise temperature to 58°F (14°C).

4. When the ferment is 75 percent complete, raise to 62°F (17°C).

5. When the ferment is 90 percent complete, raise to 66°F (19°C) and hold until terminal gravity is reached.

To better understand what we're talking about, let's use an example hypothetical lager with an OG of 1.052, which we expect to finish at about 1.010. With a 42-point gravity drop (1.052–1.010), half of that would be 21 gravity points. On to step 2: When the beer reaches 1.031 SG (52–21), it's time to raise the temperature to 58°F (14°C) and hold until the gravity drops to about 1.020. The temperature is then raised to 62°F (17°C) and held until the gravity reaches 1.014. At that point the temperature is raised to 66°F (19°C) and held until fermentation is complete.

The first part (50 percent gravity drop) can be as short as 4 to 5 days (higher gravity worts will take longer). The next steps can each occur after a day or 2. The important thing is to check the gravity—no guessing.

If you look around, you'll see some alternative fast lager ideas floating around. Brülosohphy founder Marshall Schott has a hybrid method that is more similar to methods described by Ludwig Narziss in the book *Abriss der Bierbrauerei (Overview of Beer Brewing)* and by Greg Noonan in *New Brewing Lager Beer* and incorporates a traditional lager cold-crash (http://brulosophy.com/methods/lager-method/).

Marshall waits for the beer to hit 50 percent attenuation and then ramps the temperature a steady 5°F (3°C) every 12 hours until he reaches 65 to 68°F (18 to 20°C). Once the beer is stable and without diacetyl (butter) or acetaldehyde (green apple), Marshall chills the beer 5 to 8°F (3 to 4°C) every 12 hours until it reaches 32°F (0°C). Once there, he lets it hang for 2 to 3 days to clarify.

The cool fermentation temperature of lagers and many hybrid beer styles is another source of environmental stress, slowing cell metabolism to a crawl. Add to that the stylistic requirement to minimize the ester levels in the finished beer, and we're suddenly asking a lot of our yeast. For these reasons, cold-fermented beer styles also require a higher pitching rate than beers fermented at warmer temperatures.

A good guideline when brewing high-gravity beers or lagers is to pitch double or triple the baseline pitching rate of 6 million cells/mL for ales—12 to 18 million cells/mL. And for a high-gravity lager, such as a doppelbock operating under the weight of both high OG and low temperatures, a quadrupling of the pitching rate—24 million cells/mL—wouldn't be out of order.

WHAT'S YOUR PITCHING RATE?

For those of us brewing the traditional 5-gallon (19 L) batches, low- to moderate-gravity ales can often be pitched directly from the package. But as gravity goes up and temperatures go down (or batch sizes increase), our target pitch rates will represent multiple packages' worth of cells for a batch. Alternately, we could propagate one pack of yeast in a starter culture prior to brew day to build up our pitch rate (read about making a yeast starter on page 111).

A starter culture is a great idea regardless of what you're brewing, since it ensures maximum viability and health of the yeast population, and it is an especially good practice if your yeast pack is at the older end of its date range or was purchased through mail order and shipped in hot weather or other adverse conditions.

TARGET PITCHING RATES

STYLE	GRAVITY	PITCHING TEMPERATURE	FERMENTATION TEMPERATURE	PITCH RATE (million cells/mL)
Ale	<1.060	>65°F (18°C)	>65°F (18°C)	6.00
Ale	1.061–1.076	>65°F (18°C)	>65°F (18°C)	12.00
Ale	>1.076	>65°F (18°C)	>65°F (18°C)	>18.00
Lager	<1.060	>60°F (16°C)	>60°F (16°C)	12.00
Lager	1.061–1.076	>60°F (16°C)	>60°F (16°C)	18.00
Lager	>1.076	>60°F (16°C)	>60°F (16°C)	>24.00

MORE INFORMATION

Below is just a small selection of the excellent resources for more information about yeast pitching:
- *Brew Your Own* pitching rate chart for fresh yeast: www.byo.com/resource/pitching
- Wyeast Labs online pitching rate calculator: www.wyeastlab.com/hb_pitchrate.cfm

DISCLAIMER

It's been said before, but it bears repeating: when most homebrewers talk about cell counts, it's a ballpark estimate and not an actual head count of our yeast. Pro brewers and lab types use hemacytometers and a microscope to get a more accurate count of their pitching population, but for our purposes as homebrewers, the horseshoes-and-hand-grenades approach to estimated cell counts will still yield great beer. (Using a hemacytometer is not a real head count either, by the way, but it is better than guessing!)

For homebrewing, you're pretty safe with using an estimated target range from *BYO*'s pitching chart in the link on page 110, or by punching your numbers into Mr. Malty's pitching rate calculator. (In case you didn't know, Mr. Malty is actually the co-author of *Yeast: The Practical Guide to Beer Fermentation*, which is an excellent resource for questions about yeast.)

MAKING A YEAST STARTER

Wyeast XL smack packs or PurePitch pouches of White Labs yeast contain 70 billion to 150 billion healthy yeast cells when they are packaged. After a container of yeast leaves the lab, however, its journey could be a perilous one. Factors such as temperature, transportation, and storage time all impact the viability of what eventually arrives in your hands. By the time you are ready to pitch a pack of yeast into your carefully crafted wort, can you be sure it has a sufficient quantity of live yeast as intended by the manufacturer?

To achieve their target cell count, commercial breweries usually harvest and repitch yeast from batch to batch. Most homebrewers, however, don't brew batch after batch with the same yeast strain, and so repitching is not as common at home (for more on reusing yeast, see page 113). In order to reach our cell count targets (see page 110), making a yeast starter is a good option.

Note: *One of the best ways to provide a sufficient quantity of superior yeast is building a yeast starter from a pure culture. Beginning with lab-cultured yeast provides a high probability that the strain is indeed pure. However, a pure yeast culture can only be as pure as your approach to handling it. Sanitation is essential throughout the process of creating a starter batch of yeast and cannot be overemphasized.*

Before building your microscopic yeast army, one might ask if you can have too many yeast cells in a batch of brew. While it is possible to have too much, overpitching a 5-gallon (19 L) batch of beer would require at least 400 billion cells. If you overpitch, you run the risk of producing a beer with a rubbery or excessively yeasty flavor, although the flavor differences may be small until really excessive yeast counts are reached.

Most homebrewers don't have the equipment needed to count their yeast. Instead, most rely on the fact that a yeast starter of a given size should contain a certain amount of yeast cells. For 5 gallons (19 L) of average-strength ale, a 1- to 2-quart (1 to 2 L) yeast starter should grow an adequate number of cells. For a regular-strength lager of the same batch size, a 3- to 4-quart (3 to 4 L) yeast starter is more appropriate. If you make a stirred yeast starter (see page 42), the size of your starter can be reduced by as much as half. For pitching recommendations more finely tuned to the details of your beer, see page 42.

BUILDING A STARTER

Here is how to build a yeast starter for a standard 5-gallon (19 L) batch of brew. Since it is difficult to know how many viable yeast cells survived the trip from the lab into your hands, begin by making a starter of 1 pint (about 500 mL), which you will then build up to 1 quart (about 1 L).

A yeast starter is essential for high-gravity beers. Even beers in the standard gravity range can benefit from a starter though (see below for more).

If you are using yeast in a smack pack, remove the pack from the refrigerator and allow it to reach room temperature. Then, squeeze or smack it to rupture the capsule of yeast inside and release it into the nutrients. Leave it at room temperature to incubate and swell. Allowing the pack to swell provides proof that the yeast in the pack is alive before you go any further. If you are using a pouch of yeast, simply let it reach room temperature and then gently shake the pouch to get the yeast into suspension so none of it is stuck to the bottom of it. Next, get a small batch of wort ready.

If you made a big batch of starter wort in the past and refrigerated some, grab a jar and allow it to come to room temperature. If not, make the appropriate volume of starter wort at a specific gravity of 1.020 to 1.040. (For every 1 quart/1 L of wort, use approximately 2.5 ounces/70 g of dried malt extract.) Bring it to a gentle boil for 15 minutes, cover, and let cool to 75°F (24°C). To speed the cooling process, surround the bottom and sides of the pot with cold water. Don't allow any contamination of your mini-batch of wort as you only want pure yeast culture to grow in it, and nothing else. After the wort has cooled sufficiently, open the pouch of yeast and dump it into a sanitized glass bottle or flask that accommodates at least 1 quart (1 L). (Most brewers use a 1,000 or 2,000 mL Erlenmeyer flask.)

Sanitize your hands and the outside surface of the pouch before opening it and pouring it into your container of wort. Seal the opening of the flask with aluminum foil or a foam stopper, and your yeast starter is ready to grow.

At this stage, the only thing that may be lacking in our little batch of beer is oxygen. An ample supply of oxygen is needed for the yeast to prepare for a life of fermentation. There should be some oxygen in the wort to help the yeast get started if you accompanied the transfer with splashing. You could also increase the amount of oxygen in the wort by occasionally swirling the container vigorously. Swirling also benefits the starter by keeping the yeast in suspension and in greater contact with the wort. If you have an aquarium pump or oxygen tank you use for aeration, you can also use that to more thoroughly aerate your starter.

After a day or two (at the most), the yeast in the starter should have multiplied as much as possible in the pint of wort. At this stage, make another pint-batch of sterile wort as before and feed it to the starter. Reattach the foil to the top of the flask, swirl vigorously, and set aside for another day of yeast growth. After the extra day of propagation, a substantial layer of yeast should be at the bottom of the flask. At this point, the yeast can be swirled back into suspension and the contents of the container pitched into your 5-gallon (19 L) batch of wort.

Note that the rule of thumb for propagating yeast is to increase the volume at each step by no more than tenfold. If you wanted a larger starter, you could have stepped the pint-sized (500 mL) starter all the way up to 5 quarts (about 5 L)—enough to pitch to more than 10 gallons (38 L) of ale.

If you are making a yeast starter for a beer that you know will put some strain on the yeast—for example a high-gravity or high-adjunct beer—adding some yeast nutrient to your starter may be helpful. Keep in mind that commercial yeast has already been supplied with nutrients and too much yeast nutrient can overstimulate the yeast. Check the manufacturer's recommended guidelines for addition rates and don't exceed them.

If your main batch of wort is well aerated and kept at the proper temperature for the variety of yeast pitched, fermentation should be apparent in less than 24 hours. In some cases, you may see fermentation start within 3 to 4 hours.

REUSING YEAST

Collecting, storing, and repitching yeast from batch to batch is not complicated; it just requires a little planning and special attention to sanitation. Each and every vessel, spoon, funnel, tube, or other implement that touches the yeast you plan to repitch must be absolutely clean and sanitized before use.

How you collect yeast from your fermenter will depend on the type of yeast to be collected and your fermenting equipment. If you have a fermenter that is easily accessed from the top (such as a plastic bucket) and are using ale yeast, you can simply skim some yeast off of the surface of the beer during active primary fermentation for many yeast strains. An alternative method would be to wait for the primary fermentation to subside and rinse some of the trub layer away with sterile water before collecting a sample of yeast. Then deposit the harvested yeast into a sanitized container, attach a loose-fitting lid or airlock, and place it in the refrigerator.

If you are fermenting your ale in a carboy or other vessel that is not readily accessible from the top, you will need to wait until primary fermentation is complete to harvest your yeast. In this case, rack the ale into a secondary fermenter soon after primary fermentation begins to subside and harvest the sedimented yeast from your primary fermenter. Although yeast harvested after a secondary fermentation will have less trub associated with it, the yeast from a primary fermentation will be healthier. To transfer the yeast to bottles or a keg, carefully swirl the yeast in the bottom of the fermenter and decant between a cup and a pint (8 fluid ounces or 237 mL to 16 fluid ounces or 473 mL) into a sanitized container. Attach a loose-fitting lid or airlock to the container and place it in the refrigerator. If you wish to collect yeast from a lager fermentation, follow this same procedure of collecting yeast from the bottom of the secondary fermenter.

If you are fortunate enough to have a cylindrical-conical fermenter equipped with an outlet on the bottom and a removable lid on the top, you can either harvest ale yeast from the top during active fermentation or draw off ale or lager yeast from the bottom after primary fermentation

Top cropping yeast can be harvested easily. It's shown here in a carboy, but it is much easier to harvest it from a bucket.

subsides. When accessing yeast from the bottom of a fermenter, do your best to collect the middle yeast. The sediment at the very bottom often contains dead yeast and trub, and the middle layer contains the best yeast for repitching. The bottom layer of dead yeast and trub can be distinguished by its darker color, while the preferred layer of yeast will appear more yellow and putty-like. Harvest the yeast as soon as fermentation is complete and the yeast has settled.

Now that you have a sample of yeast collected, it must be stored properly to assure continued viability. Store the yeast in the refrigerator as close to 32°F (0°C) as practical to keep it dormant. This will reduce both the risk of spoilage and of autolysis. Autolysis is when a (yeast) cell goes into self-destruct mode and essentially digests itself. This process often produces a rubbery stench that, needless to say, is undesirable in beer. As mentioned earlier, the yeast should be stored in a container with a loose-fitting lid or an airlock. After you collect the yeast and put it in the refrigerator, it will not immediately go dormant as most refrigerators are not very cold and the temperature fluctuates considerably over time. Because of this, yeast stored in a refrigerator may produce some carbon dioxide that could cause problems.

To store the yeast at a lower, more stable temperature, place the yeast container in a small cooler (or a box with baggies of ice in it). It is important to vent the container daily, for the first 3 days, because excessive CO_2 will damage yeast quickly. Likewise, some containers may rupture under excess CO_2 pressure. Yeast should be used in a batch of beer within 2 weeks of being collected from a previous batch. Yeast that is not actively fermenting can lose its viability rather quickly when stored using the simple methods and conditions outlined here. This is where scheduling to brew successive batches of beer is important if you wish to collect and repitch yeast.

When you are ready to retrieve your stored yeast from the fridge and pitch it into a new batch of wort, it is important to give the yeast a wakeup call before putting them to work. Remove the container with the stored yeast from the refrigerator and allow it to slowly warm up to the temperature of the wort it will be pitched into. You may want to give the yeast a shot of oxygen to help them become active. This can be accomplished by simply stirring the yeast up with a sanitized spoon or whisk to develop a bit of froth. If you have a magnetic stir plate, you can sanitize the stir bar, slide it into the yeast vessel, and place it on the stir plate and fire it up. If you have an air pump with an inline sanitary air filter, you could also pump air into your yeast in addition to giving it a stir to provide aeration. If your wort is adequately aerated, however, you don't need to worry too much about aerating the yeast sample. Once the yeast has been stirred and had a chance to warm, it can be pitched directly into the waiting wort.

A few final considerations for harvesting and reusing yeast relate to beer style. It is best to brew successive batches of beer that are within a fairly narrow style range of original gravity and bitterness if you plan to reuse yeast. Therefore, plan your batches accordingly so the yeast will be appropriate for each style without any undesirable flavor carryover. If you would like to use harvested yeast for a particularly dark or bitter style of beer, make that beer the last batch in the series so you can dead-end the yeast and not have to be concerned with any carryover of flavors.

COLLECTING YEAST FROM BOTTLES

A wide variety of brewer's yeasts are available to homebrewers these days. But sometimes the particular strain you want isn't commercially available. Don't give up yet: it might be possible to culture it from a bottle-conditioned beer.

Most commercial beers are filtered, and some are flash pasteurized, before bottling and do not contain yeast. However, some brewers bottle-condition some of their beers. Often, the brewer will advertise this fact on the label of those products. If not, the telltale layer of sediment on the bottom of the bottle indicates a bottle-conditioned beer.

Keep in mind, however, that some brewers use a different strain of yeast for bottle conditioning than they do for primary fermentation. The yeast on the bottom of most Bavarian hefeweizens, for example, is a standard lager strain. Franziskaner, for example, is bottled with a bottling strain, not a hefeweizen strain. One exception to this rule is Schneider Weisse, which evidence suggests is bottled with its fermentation strain. British bottle-conditioned beers, more often than not, are conditioned with their fermentation strain. To give one example, Fuller's 1845 reputedly is conditioned with its fermentation strain.

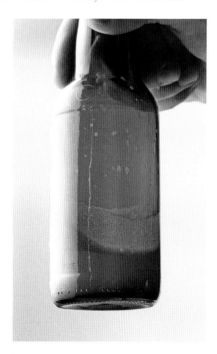

Also, you should know that some brewers use more than one strain of yeast during fermentation. If you culture yeast from a bottle, there is no guarantee that you will raise all the relevant strains. And if you do, they will most likely not be in the proportions the brewer uses. Some sources claim Saison Dupont is fermented with three strains, but—with some work—all three can be cultured from a bottle of the beer.

The success or failure of your attempt will mostly depend on the condition of the yeast you try to culture. If you get fresh beer that has been stored cold and is not extremely alcoholic, you have a good shot at recovering some yeast from the bottle. Yeast that have been stressed   from heat, age, or high alcohol levels—are much harder to culture. If you find a bottle-conditioned beer that you think is a good candidate and you wish to culture yeast from it, here's how to do it.

If a commercial beer is unfiltered, you may be able to culture the yeast at home.

PREPARING THE WORT

Before we begin, remember that keeping all the equipment you use clean and sanitized (or preferably sterile) is the most important key to culturing bottled yeast. The number of viable yeast cells in a bottle may be very low, and any contaminating microorganism that lands in your culture may grow faster than your yeast and render your culture useless. Take your cleaning and sanitation more seriously than you would for a regular batch of beer.

When culturing yeast from a bottle, the main idea is to get the yeast—which may initially be in poor health—viable and then grow it up to usable numbers. The first step is to feed the yeast some wort to "wake it up." The easiest way to do this is to add a small amount of wort to the bottle.

The wort you add should provide carbohydrates, oxygen, and other nutrients to the possibly fragile yeast cells. The concentration of these elements should be low enough not to put any stress on the yeast. Therefore, make your initial wort at a specific gravity of 1.015 to 1.020. Aerate or oxygenate it well and add a pinch of yeast nutrients. Use complete yeast nutrients, not DAP (diammonium phosphate), which is sold at many winemaking shops as yeast nutrient.

You will only need a small volume (usually 2 to 3 mL) of this wort—just enough to barely cover the bottom of the bottle. If you are culturing yeast from a 750 mL bottle, as many Belgian beers are packaged in, the amount of wort you use should be proportionally larger, around 6 mL.

At each step along the way, you want to feed the yeast a manageable amount of wort. In the early stages of culturing yeast from a bottle, sticking to 5× step-ups is a good idea. This both gives the yeast a manageable amount of wort to consume and limits the amount a stray contaminant can grow. In the bottle, aim to add roughly the same volume of wort as the sediment occupies.

WORT MEETS YEAST

Before opening the bottle, wipe the neck and cap off, preferably with a 70 percent ethanol solution or drugstore isopropanol. Pour off the beer, leaving the yeast sediment undisturbed. Use a butane lighter to flame the top of the bottle. You can also use a gas kitchen stove as a flame source. Let the alcohol evaporate before flaming the bottle and keep the alcohol capped while flame is present. Wave the bottle through the flame slowly a couple times; you don't need to torch the glass until it's hot. This should kill any microorganisms that were hiding up under the lip of the cap.

Set the bottle down and let it warm to room temperature, with the top covered with sanitized aluminum foil to prevent contamination.

Pour the feeder wort down the side of the bottle onto the yeast. If the bottle has a punt or other indentation, swirl the wort gently to rinse any yeast from the "island" into the "moat" of wort around it. Do not add enough wort to cover the "island." Flame the top of the bottle and either affix a (sanitized) fermentation lock or cover it with sanitized aluminum foil.

Depending on how much yeast is viable, the amount of time it takes to wake them up varies. Likewise, you may not always see visible signs of fermentation at the first stages of yeast culturing. The best thing to do is let the bottle incubate at relatively warm temperatures—70 to 90°F (21 to 32°C)—for 1 to 3 days, looking for foaming or a slight clouding of the wort. (Many times the top of a refrigerator will be warmer than room temperature and is a good spot for incubating cultures.)

Incidentally, don't expose the yeast culture to strong light sources either to keep it warm or to look for signs of fermentation. You don't need to keep your culture in absolute darkness, but keep it out of sunlight or other bright light.

THE FIRST STEP-UP

Once you see signs of fermentation, or 3 days have passed, transfer the fermenting wort to a small culture of fresh wort, around 15 mL. A great container for this is a 15 mL sterile culture tube, available at scientific supply shops. These come individually packed and sterilized and have a cap that can be placed loosely on the tube, allowing gas to escape during fermentation.

You can take a small cardboard box and cut a round hole in it to serve as a tube holder. This is especially handy if you are handling many tubes.

For your 15 mL culture, you may want to add a tiny pinch of lysozyme—just one or two crystals—to your wort. Lysozyme is an enzyme that kills lactic acid bacteria and is available at most home winemaking shops. Although it only kills certain strains of bacteria, this affords a bit of extra protection against contaminants that may have been present in the bottle.

To transfer the wort to your 15 mL (about half a fluid ounce) culture, remove the fermentation lock or foil from the bottle, flame the tip, let the glass cool for a second, and pour the yeast sample from the bottle to the 15 mL culture tube. Cap immediately and again incubate the 15 mL tube at 70 to 90°F (21 to 32°C).

In 1 to 3 days, you should see some sign of fermentation. This may be foam on the top of the wort, cloudiness in the wort, or just a layer of yeast at the bottom of the tube. If you see this, you are basically home-free. You have cultured yeast from your bottle. (As we'll see in a moment, this may or may not mean that it makes good beer. But, you're on your way.)

If your 15 mL (0.5-fluid ounce) culture ferments, the yeast should be healthy enough to be able to withstand a 10× step-up. To get to pitching quantities for a 5-gallon (19 L) batch, you will need two more step-ups—one to around 150 mL (5 fluid ounces) and a second to around 1,500 mL (51 fluid ounces).

THE SECOND STEP-UP

For your 150 mL (5 fluid ounces) culture, you can use wort with a specific gravity of 1.030 to 1.035. Aerate well, and maybe add a tiny pinch of yeast nutrients, but don't bother with the lysozyme at subsequent stages unless the culture smells sour.

Incubate the culture at 70 to 80°F (21 to 27°C) for 2 to 5 days, or until fermentation is complete. (The decrease in the upper temperature limit is simply to raise the yeast in the final step-ups at conditions somewhat closer to your actual fermentation conditions.)

ASSESSING THE YEAST

This is a great stage to decide whether the yeast culture you've raised will be suitable for making beer. Pitched with a 15 mL (0.5 fluid ounces) culture, a 150 mL (5 fluid ounces) culture should ferment like a little batch of beer. You should see the kräusen rise and fall as the yeast works. And, when they're done, the resulting liquid in the culture should taste like beer.

To find out if this is the case, decant a decent amount of the liquid into a sanitized container and refrigerate overnight. A 50 mL culture tube works great for this, but a baby food jar—or any other small, sealable container—will work just fine. Fill the container to the rim, seal, and store cold. Refrigeration will settle most of the remaining yeast out of the liquid and allow you to taste the starter beer without a yeast bite interfering with your sampling. If the starter beer tastes good, then step the beer up to 1,500 mL (51 fluid ounces) and brew with it. If it doesn't, chalk it up to experience and try again.

THE FINAL STEP-UP

For the final step-up prior to pitching, there is one small twist on the procedure. Let the 150 mL (5 fluid ounces) culture ferment completely and then only pitch the yeast sediment. In the previous steps, you pitch the whole culture. The reason for pitching only the sediment is to discard any mutant cells (if present) that are not flocculant. Once the 1,500 mL (51 fluid ounces) yeast starter shows activity, it is ready to pitch. Alternately, you can let it ferment out and perform a second taste before using the yeast on a whole batch of precious homebrew.

WHAT ABOUT LAMBICS?

You can culture bacteria and wild yeast from beers as well as brewer's yeast. Lambics are a popular target for homebrewers to grab cultures from. Bottle-conditioned lambics, such as those from Cantillon, have a wide variety of microorganisms potentially present. By using the step-up procedure—minus the lysozyme, of course—you will most likely obtain a mixed culture of microbes.

Keep in mind that some of these microbes have growth conditions that are dissimilar from brewer's yeast. Growing up *Brettanomyces*, *Lactobaccilus*, or *Pediococcus* takes more time than raising brewer's yeast does. As a result, many home lambic brewers simply maintain a standing mixed culture of "bugs" they have harvested from various bottles. You can also purchase a lambic yeast blend from a commercial yeast company rather than go through the lengthy process of raising up the bugs from scratch.

WATER

Water is the main ingredient of beer. The many different styles of beer we have today evolved for many different reasons, not the least of which is the chemistry of the local water supply where the beer was created. The details of water chemistry can be complex, and many brewers may simply wish to know if they can use their water as is or learn a simple treatment plan to deal with their water.

THE SHORT COURSE ON WATER TREATMENT

Historically, brewers no doubt experimented with different ingredients and techniques much as homebrewers do today. They undoubtedly settled on recipes that worked best for what they had readily available, including water. Without having a comprehensive knowledge of what was dissolved in their brewing water and its effect on mash pH, expression of hop bitterness, etc., the brewers nevertheless found their way, and the rest, as they say, is history.

Two things all brewing waters require are that they taste good and be free of chlorine compounds. You can easily do this two ways: either buy distilled or RO (reverse osmosis) water, or remove all chlorine/chloramine from your tap water with a water filter. There are other ways to treat for these, which we'll cover later in this section, but filtering or using RO or distilled water are the simplest methods.

If you are an extract brewer, the requirements your water must meet are broader than if you are an all-grain brewer. This is because you do not have to worry about mashing your grains. This has been done for you at the malt extract plant. Malt extract is condensed wort, and all (or most) of the dissolved solids present when the grains were mashed are contained in the extract, including mineral ions. When brewing with malt extract, you simply reconstitute your wort from this concentrate. If you use distilled water, or very soft tap water, your reconstituted wort should contain all the minerals required for brewing. If your water is hard, you will be adding minerals to your wort beyond what is required. In small amounts, this will likely have no discernible effect. If your water is very hard, especially if it is rich in carbonates, you may want to consider blending it with distilled or RO water to make your brewing water.

You do not need to add salts to your brewing water to try to emulate the water of different brewing cities (such as Burton). Unless you know the minerals that are already in your malt extract, you are blindly piling on more minerals. For hoppy beers, you may wish to accentuate the hop profile by adding a little gypsum. Likewise, for malty beers, a little calcium chloride may make for a smoother beer. In either case, don't overdo the addition. Use a maximum of 2 teaspoons per 5 gallons (19 L) of these salts.

All-grain brewing additionally requires that the water chemistry yields a suitable mash pH. A simplified version of how to obtain this can be had by remembering a few key things. Calcium ions (and to a lesser extent, magnesium ions), dark malts, and acids will lower mash pH. Carbonates neutralize acids and decrease the amount that mash pH is lowered by these things. All-grain brewers should test to make sure your water pH is within the ideal range if you're using tap or well water. Simple water chemistry calculators are available online that will do all the math for you. These include Greg Noonan's "Water Witch," available for download at https://byo.com/resource/brew-water-spreadsheet/, as well as others such as EZWaterCalculator (www.ezwatercalculator.com) to adjust your pH if it's too high or low.

Calcium has other beneficial actions in brewing, such as stabilizing alpha amylase in the mash. Thus, unless stylistically required to have less, it's best to have at least 50 ppm calcium in your wort. For pale beers, the amount of carbonates should be minimized, at least under 50 ppm. For dark beers, carbonate can be a good thing, and for stouts, your water may require up to 250 ppm carbonates.

Water adjustment chemicals may not look like much, but they can take a beer from good to great.

TIPS FROM *BYO*

CALCULATING WATER USAGE

Just as important as water chemistry–and extraordinarily practical–is the quantity of water used when brewing. Such questions as how much water to use for mashing and sparging the grain and for boiling the wort are of great interest to brewers.

If you are using water that is filtered or adjusted for mineral content, you will want to know the total volume needed for a brewing session. The volume of water also impacts brewing equipment and system design because it is a major factor in determining the capacity of the vessels required.

All of the good comprehensive brewing software packages available have a brewing session water volume calculator that performs this task for you. However, even if you rely on the computer to do the routine calculations, an understanding of the underlying principles will make you a more knowledgeable brewer and might even improve your beer.

When calculating water quantity, it is helpful to begin by considering the final volume of beer you intend to brew. Recipes are usually expressed in round numbers– for example, 5 gallons (19 L)–and often they reflect the volume that is bottled or kegged. During fermentation, racking to other vessels, and the bottling or kegging process, some beer is absorbed by hops or yeast, while an additional volume is evaporated or left behind in the equipment. After some experience, brewers come

to know approximately how much wort is necessary in the fermenter in order to end up with a given volume of beer. I call this the *fermenter volume*, and to my mind, this is the real volume of a recipe. For 5- to 10-gallon (19 to 38 L) batches, typically this will be 1 to 4 quarts (0.9 to 3.8 L) larger than the published volume of the recipe.

Once experience has taught you the desired fermenter volume, you can work backward to determine the total water volume needed, subtracting the various losses that occur during the brewing process. In reverse order, these will include many of the following:

- Wort left in lines and equipment, such as a chiller or pump, between the brewing kettle and the fermenter
- Wort absorbed by hop residue and protein break material in the kettle at the end of the boil
- Evaporation losses during boiling of the wort in the kettle
- Liquid left in lines and equipment between the mash tun and the kettle
- Mash "dead space," that is, liquid that is left in mashing and sparging vessels due to their design and geometry
- Sparge water dead space, that is, water similarly left in the hot liquor tank or other sparge water vessel
- Water absorbed by the grain in the mash tun

Of course these losses are offset by water additions, which typically will include at least some of the following, listed in order of earliest to latest:

- Strike water, that is, the water initially mixed with the grain at the beginning of mashing
- Any additional water infusions during mashing
- Sparge water added to the mash in order to extract the sugars converted from the starches in the grain
- Any water added to the kettle to achieve the target preboil wort volume
- Any water added after the boil, either to the kettle or the fermenter, to achieve the target post-boil and fermenter volumes

Some of these values can be calculated from the recipe and brewing method, while a number of them are derived from the relationships among values, and still others are measured empirically based on the specifics of the equipment and brewing system. For much more detail on this topic, visit https://byo.com/article/calculating-water-usage/.

TOP: Filtering to remove chlorine and other impurities is the most important thing you can do when it comes to your brewing water.

BOTTOM: A pH meter is a useful tool. If you don't want to invest in your own, see if anyone in your club has one you can borrow.

Manipulating just the calcium and carbonate levels is the simplest way for an all-grain brewer to treat their water. To reduce carbonates, if needed, you can blend your tap water with distilled or RO water, or add acid. If you need to increase it, you can add calcium carbonate (chalk) or sodium bicarbonate (baking soda).

Once your carbonate levels are adjusted, you can add calcium—if needed—as either calcium sulfate (usually in the form of gypsum) or calcium chloride ($CaCl_2$).

Whether an all-grain or extract brewer, always taste your brewing water after you have treated it, and don't proceed if you detect off-flavors or aromas. If you do taste something off, double-check that you used the correct mineral salts. Don't taste your water for several hours after adding Campden tablets.

WHAT'S IN THE WATER?

With the information and technology we have today, we can discern what ions are present in solution in our water and the water of famous brewing regions of the world. If you live in the United States and use municipal water, your local water board should send you a summary of what is in your water every year. If you don't receive this publication, you can request it. Alternately, if you have a private water supply (well, spring, etc.), you should have it analyzed by a private or state-run lab that does such work.

Results shown on water analysis reports are typically expressed in milligrams per liter (mg/L) or parts per million (ppm). In the range of concentrations we are concerned with, you can use these interchangeably. Once you have an analysis of your brewing water, you can compare it to the analysis of the water from brewing centers from around the world and see which beer style fits your local water best. Before we dive into that, though, here's a quick refresher on the basics of water chemistry.

Water is a molecule composed of a central oxygen atom with two smaller hydrogen atoms attached. A space-filling model of water looks a bit like Mickey Mouse, with the oxygen being his head and the two hydrogens his ears.

Many different types of minerals dissolve in water. When they do, some dissociate (break apart) into their component ions. For example, sodium chloride (NaCl) dissolved in water would dissociate into two ions, $Na+$ and $Cl-$. An ion is simply an atom or molecule that has a different number of electrons than protons. This difference results in either a net positive charge (cation) or a net negative charge (anion). These ions float around in solution in water and are available to react with other ions and affect everything from mash pH to the flocculation of yeast. What if your water is best suited for a Pilsner and you wish to brew an Irish stout? Various salts may be added to adjust your water chemistry to suit a particular style of beer, but first let's look at some important ions in brewing water, how they affect beer flavor, and how their concentrations relative to each other can create flavor synergies.

BASIC BREWING CHEMISTRY

Pure water—for example, distilled water or water purified by reverse osmosis (RO)—without any mineral ions in solution is not used by commercial brewers. It is the dissolved ions in water that are important for mash chemistry, expression of various flavors (sweet, sour, salty, and bitter in particular), and yeast nutrition. Therefore, it is not a good idea to brew all-grain beer with distilled or reverse osmosis water.

Water with high levels of minerals dissolved in it, especially calcium ($Ca2+$) and magnesium, is called hard water. Water with few dissolved minerals is called soft water. For brewers, it is more important to know the concentrations of key minerals that are dissolved in their brewing liquor than whether their water is hard or soft.

Water with very low levels of carbonates (e.g., in Pilsen, Czech Republic) will allow mash pH to come into proper range (5.2 to 5.6) with only pale malt (especially if a little calcium is present). If the carbonate levels are higher, more acidic malt is necessary to properly lower the pH. For example, hard water dominated by carbonates such as in Dublin, Ireland, with its high level of carbonates, is well suited to brewing stouts. Carbonates and bicarbonates can be precipitated as calcium carbonate ($CaCO_3$) by boiling water in an open kettle for at least 15 minutes where it can pick up oxygen to react with and drop out of solution. This process will typically reduce carbonates and/or bicarbonates below 150 ppm. Carbonates can also be reduced by neutralizing them with acid. Food-grade phosphoric acid is a popular choice for this in breweries.

Calcium ions (Ca^{2+}) in water react with phosphates in malt, releasing acid. Thus, its presence in mash water lowers the pH of the mash. Calcium is not a significant yeast nutrient, but it does facilitate yeast flocculation and subsequent precipitation. Calcium also stabilizes alpha amylase and increases its tolerance to the heat of mashing.

Magnesium (Mg^{2+}) is important for enzyme activity in the mash and for yeast nutrition. Like calcium, magnesium ions drive down the pH of a mash, but to a much lesser extent than calcium. Magnesium enhances beer flavor up to a point, then lends a dry, bitter metallic flavor to beer.

Sodium ($Na+$) has different effects at different concentrations. At low levels, sodium contributes sweetness, probably by balancing bitterness, and adds some palate fullness, which may be appropriate in certain styles of beer. At higher levels, sodium can contribute to salty flavor.

Sulfate (SO_4^{2-}) has a very high solubility in water, and waters high in sulfate are known as gypseous waters. Sulfate ions bring out the hop character in a beer.

Chloride ions ($Cl-$) give a full, sweet flavor to beer, but they are not a significant player in mash chemistry or yeast nutrition. Many brewers use calcium chloride instead of calcium sulfate because chloride has a flavor effect that many brewers like.

Chlorine (Cl) is often included in municipal water supplies as hypochlorous acid ($HOCl$) or chloramine (NH_2Cl) to serve as a disinfectant. If either of these compounds remains in brewing water, they can lend a harsh, medicinal flavor to beer. Hypochlorous acid can be removed by boiling brewing water in an open pot for at least 15 minutes prior to using the water in the mash or boil. Remove chloramines from water by adding a crushed Campden tablet to 20 gallons (76 L) of brewing water and letting the water sit uncovered overnight to allow the resulting chlorine gas to dissipate.

Of course, as mentioned at the beginning of this section, the simplest way to lower an ion in your water is to substitute a portion of your tap water with distilled or reverse osmosis (RO) water that is essentially free of mineral ions. For example, to reduce the concentration of the ions in your tap water by half, use half tap water and half distilled or RO water to make up the total volume of water for the batch of homebrew.

REVERSE OSMOSIS WATER ADJUSTMENTS

Reverse osmosis (RO) water has very few minerals, so you can use it as a starting point for building brewing water for your purposes. Because it has very little alkalinity (dissolved carbonates and bicarbonates), it doesn't buffer acids much—this means acid additions can easily lower pH rather than having to first overcome these buffers. Since the water will be extremely low in minerals, you don't have to worry about variability in your water source or get water reports to analyze.

The mash pH for base malts generally settles at 5.8–6.0, which gives an indication of what other acidification is necessary when brewing. If you are using RO water, mashing base malts should result in a fairly consistent pH that you can measure and use as part of your recipe planning. Using a good-quality calibrated digital pH meter, you can measure how much acid you need to add to your mash to lower the pH to the desired range.

Gordon Strong acidifies all of his RO brewing water to a pH of 5.5 (at room temperature) using phosphoric acid. A portion of this water is used for the mash, and measured amounts of water salts are added directly to the mash. To make water adjustments easier, he recommends mashing only base grains and starchy adjuncts—things that actually need to be mashed. Leave out roasted and crystal malts from the mash since they are already converted and add them during wort recirculation after the mash is over. The pH during the mash thus becomes very predictable without the use of recipe software or water calculators. If you don't acidify your mash with (liquid) acids, you may find that you need to add 2–3 percent acidulated malt to your grist to achieve your desired mash pH.

Generalities about the local brewing water in classic brewing cities are relevant to inform choices. (Charts detailing specific information about the brewing water of famous brewing regions are readily available online, and you can find a few on the opposite page.) However, charts don't always tell the whole story, and it is helpful to think of beer styles by their major characteristics and then decide if you want to accentuate certain flavors through the use of flavor ions (e.g., table salt or Epsom salt). The nice thing about this approach is that the flavor ions can also be added to finished beer. So, if you aren't sure about the flavor profile you want, you can run some tests with your completed beer and decide what you want to incorporate into your next batch. Just run those tests as bench trials on a small glass and scale up when you find what you like. Here are some general guidelines when brewing with RO water:

- For most beers, use 1 teaspoon of calcium chloride per 5 gallons (19 L) of finished beer, which should provide about 60 ppm of calcium and 107 ppm of chloride. This will provide ample calcium for the mash and allow the flavors of the beer to come through cleanly.
- For some hoppy styles (like pale ales), instead use 1 teaspoon of calcium sulfate per 5 gallons (19 L) of beer. This provides about 59 ppm of calcium and 141 ppm of sulfate.
- For some more balanced styles (like Kölsch), use ½ teaspoon of both calcium chloride as well as calcium sulfate.
- West Coast–type American IPAs use a lot of sulfate, but New England–type IPAs favor chlorides. Think about the difference in how those beers finish to give you an idea of the range of adjustment.

- If making a beer style that is known for a minerally profile (like Dortmunder export or English IPA), double the level of salts—2 teaspoons per 5 gallons (19 L), using half of them in the kettle and half in the mash.
- If brewing a Czech Pilsner or other very low mineral beer, halve the standard amount— $\frac{1}{2}$ teaspoon of salts.
- Try to avoid using sulfate salts in conjunction with noble hops, because it can create a clashing flavor.

These additions assume that you have treated your RO brewing water to pH 5.5 with phosphoric acid and are adding your crystal malts and dark grains after the mash has finished. If you follow other processes, you will have to account for adjustments needed to hit your target mash pH. Getting to know what is in the water you brew with and how it can be manipulated to fit the style of beer you wish to brew can make a big difference in your finished beer. After all, water is the main ingredient in everyone's brew.

DESIRED ION CONCENTRATIONS IN BREWING WATER

	Desired Range	Minimum	Maximum
Calcium	Ca^{2+}	50	150
Magnesium	Mg^{2+}	10	30
Sodium	$Na+$	0	150
Carbonate	CO_3^{2-}	0	250
Bicarbonate	HCO_3^-	0	250
Sulfate	SO_4^{2-}	50	350
Chloride	$Cl-$	0	250

WATER PROFILES OF FAMOUS BREWING REGIONS

City	Calcium	Chloride	Carbonate or Bicarbonate	Sodium	Sulfate	Magnesium
Burton-on-Trent, UK	295	300	55	725	45	25
Dortmund, Germany	250	550	70	280	25	100
Dublin, Ireland	115	200	12	55	4	19
Edinburg, Scotland	120	225	55	140	25	20
London, England	52	156	99	77	16	60
Pilsen, Czech Republic	7	15	2	5	2	5

FRUIT AND SPICE

In addition to the four main ingredients of beer (malt, hops, yeast, water), there are many other ingredients you can add to make your homebrew unique. In this section, we'll explore the more common fruits and spices that you can use to make beer.

FRUITS FOR BREWING

Fruit beers, like many styles, have enjoyed a resurgence with the craft brewing explosion over the last few decades in the United States. Craft breweries have led the way, often blending or designing beers around the unique flavors that their favorite fruit provides.

Homebrewing beer with fruit involves a little bit of art and a bit of science. Fruit beers are frequently formulated to be light tasting, light bodied, and also lightly hopped. The reason for this is simple: most fruits lose a lot of their sweet flavor during fermentation. While most fruits have a substantial amount of simple sugars in them, most of these sugars ferment directly into alcohol. If you are adding 1 to 3 pounds (0.45 to 1.4 kg) of fruit per gallon (3.8 L), this will drive up the alcohol content of your beer significantly, depending on the sugar content of the fruit. Fruit will not add any maltiness, sweetness, or body to your beer, however—just alcohol. The flavor and aroma of fruit is also often fermented away, so if you brew a beer with a lot of fruit and not much malt in an attempt to keep the gravity under control, you will end up with a thin beer. Similarly, if you start with too much malt, you will end up with a malty beer with too much alcoholic warmth that can mask the fruit flavor. Finding the right balance between fruit, malt, and hop flavors can be a real challenge. Oftentimes crafting the perfect fruit beer recipe will take more than one attempt.

A strong malt or hop flavor can overpower the subtle fruit flavors, making the fruit undetectable in the finished beer. A wide variety of beer styles can use fruit, however—even stout! For malts, it really depends on your recipe. Though if you want the fruit to shine through, keep that in mind. You don't want a strong-roasted, smoked, or even caramel flavor competing with the fruit for most beers. For hops, it is best to go with low alpha bittering hops, often with a single boil addition. This minimizes hop aroma and flavor, which allows the fruit aroma and flavor to shine through. Noble hops are often a good choice. Use whirlpool or dry hops sparingly.

For yeast, stick with clean-finishing, high-attenuating yeasts. This is not to say you could not try a more complex yeast, but strains that are low flocculating as well as low-attenuating yeasts generally take longer to fully ferment the sugars in the fruit. The complex flavors of the yeast don't always complement the fruit flavor itself. Also, the low-flocculating yeasts create more clarity problems— which is already an issue with most fruit beers.

Keep in mind that many fruits carry some wild bacteria and wild yeast with them, which can often add some sourness or complexity to your fruit beer. This is also the reason why many fruit beers take additional time to reach full maturation.

COMMONLY USED FRUITS

Apple: Apples in beer produce only a mild flavoring. Apples are (generally) best used with meads and hard cider as they tend to be acidic in flavor and don't provide a strong profile. Apple can be added as either fruit pulp or cider/juice. If using fresh fruit, start with about 2 pounds per gallon (0.9 kg per 3.8 L) and experiment from there.

Apricot: Apricot works much better in beer than peach, and it produces a peach-like flavor in the finished beer. If you want peach flavor, use apricots at a rate of 1.5 to 4 pounds per gallon (0.6

to 1.8 kg per 3.8 L). Apricot extract also produces good results.

Blackberry: Blackberry, like raspberry, is another great fruit to use in beer. However, they do not come through as intensely as raspberry, requiring a larger usage rate of 1 to 3+ pounds per gallon (0.45 to 1.4+ kg per 3.8 L). The color also carries over well to the finished beer.

Blueberry: Blueberry does not hold up well in beer. Some brewers claim that cooked blueberry holds up better than uncooked, but fermented blueberries are a very subtle flavoring. If you want to give it a try, use 2 pounds per gallon (0.9 kg per 3.8 L).

Cherry: Traditionally used in many Belgian beers, ripe, sour cherries are best as they blend well with malt flavors. Cherries should be pitted as the seed contains cyanide compounds. Generally a lot of cherries are needed, as much as 1 to 3 pounds (0.45 to 1.3 kg) per gallon (3.8 L) of beer, which is why many cherry-based Belgian beers are expensive.

Cranberry: Cranberries add a dry tartness and color to a beer, but unfortunately do not contribute much flavor. Freeze and purée them before adding them to the secondary. Use 1.5 to 4 pounds per gallon (0.6 to 1.8 kg per 3.8 L).

Moving beyond the traditional brewing ingredients is a surefire way to get unique flavors in your beer. Just remember: less is often more!

Grapefruit: Grapefruit is excellent for pairing up with some of the grapefruity American hop varieties, and you can add it as either zest or fresh fruit/juice, just beware of its characteristic bitterness and avoid the white pith of the peel. Start experimenting with 1 ounce (28 g) of zest or one to three grapefruits' worth of juice per 5 gallons (19 L) and experiment from there to your taste.

Lemon or lime: Both of these citrus fruits have very strong flavor additions that are acidic as well. These should be used sparingly as they can easily overpower the flavor of a beer. Start by trying 1 ounce (28 g) of zest or the juice of ten or so lemons or limes per 5 gallons (19 L).

Peach: Peach is a fruit that fades when used in beer, though its sugar will add alcohol. Apricot is a good substitute that creates a flavor similar to peach in the finished beer. Peach flavoring is also a possibility. Use 1.5 to 4 pounds fresh peaches per gallon (0.6 to 1.8 kg per 3.8 L).

Pear: Like apples, pears are more widely used in ciders and meads. They only provide a subtle flavor to the beer, but they can be a refreshing addition. Like apples, pear juice/perry can be added or you can use fresh fruit. If using fresh fruit, target about 2 pounds per gallon (0.9 kg per 3.8 L).

Pineapple: This tropical fruit provides a very subtle acidic flavoring. It requires 2+ pounds per gallon (0.9+ kg per 3.8 L) to generate any significant flavor in the finished beer.

HARD SELTZER

The hard seltzer craze took off around 2019. They appealed to many consumers for being lighter on the palate, skinnier on the calories, and less challenging to consume than big-bodied, highly flavorful, calorically rich beers. Macrobreweries and craft breweries of all sizes jumped on the trend as it took a big bite into the market share and shelf space dedicated to beer. Homebrewers quickly followed suit.

While homebrewers can replicate these hard beverages in a way that American commercial breweries can't, thanks to the US Alcohol and Tobacco Tax and Trade Bureau, by adding vodka, citric acid, and flavorings to carbonated water, we'll focus here on creating fermented hard seltzer, which is a bit trickier.

At its most basic, this approach boils down to fermenting sugar water into what the tax collector defines as beer. The problem is that sugar water lacks the nutrients that make wort an ideal solution to ferment into beer. Sure, sugar water can be fermented by yeast to produce a crude mixture of water and alcohol, but the resulting flavor is not clean and is not the sort of product that can be enjoyably consumed without further processing. This is where yeast nutrients enter the picture as a crucial ingredient required for the production of any fermented beverage using a high proportion of simple sugars. The term "yeast assimilable nitrogen," or YAN (where YAN = nitrogen from ammonia + nitrogen from amino acids), is most often used by winemakers and meadmakers to quantify the amount of nitrogen compounds in must–for example, grape juice–that can be metabolized by yeast. By knowing the YAN concentration of must, or sugar water in the case of hard seltzers, nutrient adjustments can be made by referring to tables relating nutrient addition scheme to YAN concentration.

Must containing less than 50 ppm of YAN is a high-risk environment for yeast and a combination of organic nitrogen (amino acids), inorganic nitrogen (ammonia), and micronutrients, such as zinc, manganese, and B vitamins, is recommended for healthy fermentations. The most common nutrient blends are proprietary and contain varying concentrations of lysed yeast cells and diammonium phosphate (DAP). Lysed yeast cells provide amino acids and micronutrients, and DAP is a ready source of ammonia. To make things easy, there are a number of yeast and nutrient blends specifically developed to cleanly ferment seltzers. A clean-fermenting brewer's yeast, Champagne yeast, or distiller's yeast are all commonly used to ferment seltzers.

Another difference between fermenting hard seltzer versus malt beverages is that the sugar used to make seltzer is completely fermentable by yeast, which means that a fully attenuated hard seltzer base will have little to no fermentables left at the end of fermentation. Knowing this, the starting gravity in seltzer is lower than a beer with the same resulting ABV.

Fruit flavors are one of the appeals of seltzer. The easiest way to flavor a neutral hard seltzer base is by adding unfermentable flavorings and organic acids to the finished and carbonated base. Another simple way to make variants from a single batch of neutral base is to add flavored syrups, a la Berliner weisse, immediately before serving. This is especially appealing for brewers who don't want to tie up multiple taps for seltzer.

The recipe and instructions provided here put this information to use to create a citrus-flavored hard seltzer. It's a good beginner's recipe, which can then be adapted to include different flavorings, yeast and nutrient combinations, and beyond.

HARD SELTZER

(5.25 gallons/20 L) OG = ~1.031 FG = ~1.000 ABV = ~ 5%

NEUTRAL BASE INGREDIENTS

5.5 gal. (21 L) reverse osmosis (RO) or distilled water
4.5 g gypsum (adjusts calcium concentration to 50 ppm)
3 lb. 4 oz. (1.65 kg) cane sugar
11.5 g packet SafAle US-05 or other neutral yeast strain
2.5 g Yeastex 82 yeast nutrient (added with yeast pitch)
2.5 g diammonium phosphate (DAP) (added with yeast pitch)
2.5 g Yeastex 82 yeast nutrient (added 36–48 hours after yeast pitch)
2.5 g diammonium phosphate (DAP) (added 36–48 hours after yeast pitch)

FLAVORINGS

2.8 ml lime extract added after CO_2 bubbling (if used) and carbonation
2.8 ml lemon extract after CO_2 bubbling (if used) and carbonation
420 ml pulp-free orange juice (single-strength, not from concentrate) after CO_2 bubbling (if used) and carbonation
14 g citric acid powder after CO_2 bubbling (if used) and carbonation

STEP BY STEP

Add water and sugar to kettle, turn on heat, and stir to dissolve sugar. Check solution strength and adjust as necessary; the preboil gravity should be ~1.031 OG. Continue heating until solution is boiling and boil for 20 minutes. Check solution strength and dilute as necessary with RO or distilled water. Cool to 64°F (18°C) and transfer to fermenter. Add yeast and first addition of nutrients.

Fermentation should begin within about 12 hours. The second nutrient addition is added once fermentation has really kicked off and the gravity has dropped by about 0.008 OG; this should be about 36–48 hours after yeast pitching. When fermentation is complete (5–7 days after yeast pitch–FG will depend on yeast and nutrients but should be around 1.000), cool to 32°F (0°C) (or as cold as possible at home) and hold cold to permit yeast sedimentation.

Next comes clarification, if you choose to clarify. Commercially produced seltzers are clarified by filtration and/or centrifugation, but hard seltzer is easy to clarify using a cartridge filter home. Then transfer the seltzer into a keg and carbonate to 2.8–3.0 volumes. Bottle or keg conditioning is not typical for hard seltzer. Taste the seltzer base to determine if aroma stripping is needed.

Sulfur off-flavors are fairly common with hard seltzers. Carbon dioxide bubbling can be used to strip these unpleasant rotten-egg and burnt-match aromas from hard seltzer. Just make sure to vent the keg during bubbling to allow these volatiles to escape.

The last step is to add the flavor additions. Dissolve the citric acid powder in the orange juice and add it along with the lemon and lime aromas. Slowly release the pressure on the keg, open the top while flushing the headspace with CO_2, and add the flavorings. Quickly closing the lid and repressurizing the headspace will minimize loss of carbonation. The orange juice in this recipe is intended to provide a slight haze, so if you choose not to filter, the juice will complement the haze of the seltzer.

Some fruits have such a high water content that it takes a large amount to make a small impact. Other fruits, such as citrus, can make a big impact with a small amount of the fruit's rind.

Plum: Plums are a great addition to a variety of beer styles, including many darker styles. Use 0.5 to 2 pounds per gallon (0.23 to 0.9 kg per 3.8 L).

Raspberry: Raspberry is one of the best fruits for brewing. The flavor and aroma hold up well to fermentation and come through well in the finished beer. The flavor is strong even at a rate of 0.5 to 1 pound per gallon (0.23 to 0.45 kg per 3.8 L), making raspberry a favorite fruit choice of brewers.

Strawberry: Strawberry is a tough choice for brewing. The flavor and color fade quickly in beer, and the aroma is very subtle. If you are going to use strawberry in a beer, you need to use fully ripe berries, you must use a lot of them (2 to 5 pounds/.09 to 2.3 kg per gallon), and you must drink the beer as young as possible as the flavor and aroma will be gone before you know it.

Watermelon: Watermelon provides a subtle creamy flavor to beer, though it takes quite a bit of watermelon to get any pronounced flavor. Ripe or overripe watermelon works best, and you can use as much as 3 to 5 pounds per gallon (1.4 to 2.3 kg per 3.8 L). Adding watermelon rind to the beer will sour it.

Other fruits: A variety of other fruits are less commonly used in beers and meads, such as dates, bananas, mangos, pomegranate, etc. Most of these fruits produce only a mild flavor and aroma, though they add considerable fermentable sugars.

HOMEBREWING WITH FRUITS

Freeze whole fruit once and thaw it before adding it to the beer. Freezing fruit breaks open the cell walls, allowing more flavor and aroma to permeate the beer. Thaw the fruit before use and bring it up to room temperature before adding it to your beer, however, to avoid shocking the yeast.

Add the fruit to the secondary fermenter if at all possible. Since whole fruit in particular contains a lot of microbes and bacteria, adding fruit too early in the fermentation process can lead to infection. By the time your beer is in the secondary fermenter, it has a higher alcoholic content, is more acidic, and also is nutrient depleted but yeast rich, all of which serve as a guard against potential infection.

One cautionary note when working with glass carboys as a secondary fermenter: adding fruit to your beer will cause rapid and vigorous fermentation, which requires several gallons or liters of headspace above the beer. Be sure you have adequate headspace and ventilation in your fermenter to prevent the bubbling trub from blocking your airlock, which could make a bomb out of a glass carboy.

Juices, concentrates, and aseptic fruit purées can also be used much like whole fruit—add them directly to the secondary. Adjustments must be made for concentration, however, as concentrated fruit juice contains more flavor and fermentables than natural juice.

Beer clarity can be a significant problem when brewing with fruits. Most fruits contain pectins, carbohydrates, and proteins that contribute to haze or cloudiness in the finished beer. If you boil your fruit, in particular, you may see a pectic haze unless you use a pectic enzyme to reduce it (for more on using pectic enzyme, visit https://byo.com/mr-wizard/pectic-enzymes/). You can also use a fining agent when brewing with fruit, and best results may be achieved using a combination of methods to achieve better clarity.

Aging is another issue when working with fruit beers. Fruits contain many wild yeasts that can secrete enzymes that break down malt dextrins and lead to more fermentation. For bottle-conditioned beers, this can be a significant problem as the bottle that was perfectly carbonated a month or two after bottling may be an overcarbonated gusher a month or two later.

The second aging issue is that the flavor profile of fruit beer will inevitably change over time. Young fruit beers may have a poor flavor profile due to unfermentables, as well as the pectins, proteins, and other complex fruit materials in the beer. At some point the flavor of the beer will definitely peak, but for some fruit beers this can take 6 months to a year or more. Finally, as the fruit continues to change, you may see a drop in quality once the beer is past its peak.

Many brewers also blend fruit beers with other beers or fruit extract before bottling. Many Belgian styles such as lambic are blended beers using a combination of well-aged and younger brews. Craft breweries often blend fruit-heavy beers with a lighter beer to balance the flavor. It's also not uncommon to add fruit extract to a fruit beer to either enhance the fruit flavor or complement it by adding another flavor.

SPICES FOR BREWING

If you decide to delve into the world of spices in beer, the first thing you should consider are any spicy flavors that may be present in beer. These flavors can come from malt, specialty grains, and yeast by-products, and you must think about how they will play with any additional spices. Due to this, and due to the wide variety of different flavors and aromas found in various beer styles, it is difficult to give precise quantities and situations for the use of each and every spice. In the end, it depends on the particular flavor and aroma profile you seek in your beer. This is certainly an area of exploration where brewing is a bit more art than science. However, in this section we will explore some ideas on how to approach spicing beers as you work toward formulating your own signature spiced homebrew.

When considering a spiced beer, it is best to begin with a final flavor profile in mind. Also keep in mind that less is more when it comes to spices in beer. Spice additions for a typical 5-gallon (19 L) batch of beer can be as small as 0.1 ounce (3 g) of a dried spice. If you don't have a sensitive enough scale to weigh out small quantities of spices, it may be easier to measure a volume such as 10 teaspoons of a spice and record the weight. For example, if 10 teaspoons weighs 1 ounce (28 g), then you can use 1 teaspoon to obtain 0.1 ounce (3 g).

When taking a sip, you want your spices to blend with the flavor profile of your beer, not bomb the other flavors into submission (well, unless that truly is your goal). In other words, someone drinking your beer should ask "Is there a bit of cinnamon in this?" rather than, "Whoa, how much cinnamon did you put in this?!" With all that said, on the opposite page is a chart of the most common spices used in brewing, and when and how they are typically used.

Using spices such as cinnamon and vanilla can create a wonderfully unique stout.

Ingredient	Part of plant used (dried unless otherwise noted)	Maximum amount in a 5-gallon batch	When added to brewing process	How long? (minutes unless otherwise noted)
Allspice (*Calycanthus floridus*)	Seed, whole	.2 oz.	Boil	45
Bitter Orange Peel (*Aurantium amarae pericarpium*)	Peel	1 oz.	Boil	15
Cacao (*Theobroma cacao*)	Bean ("nibs")	8 oz.	Secondary fermenter	10 days
Chile (*Capsicum annuum*)	Pod, fresh	.25 lb.	Steep	15
Cinnamon (*Cinamomum zeylanicum*)	Bark	4 sticks	Boil	30
Cloves (*Eugenia aromatic*)	Bud, whole	10 buds	Boil	30
Coffee (*Coffea arabica*)	Bean, ground and extracted in water	12 shots espresso	At kegging or bottling	n/a
Coriander (*Coriandrum sativum*)	Seed, crushed	2 oz.	Boil	15
Elderberry (*Sambucus nigra*)	Flower	2 oz.	Secondary fermenter	2 days
Ginger (*Zingiber officinale*)	Root, fresh grated	6 oz.	Boil	15
Grains of Paradise (*Aframomum melegueta*)	Seed, ground	.1 oz.	Boil	5
Juniper (*Juniperus communis*)	Leaf	4 oz.	Boil	60
Lavender (*Lavandula angustifolia*)	Flower	1 oz.	Steep	15
Lemon (*Citrus limon*)	Zest, fresh	1 oz.	Steep	15
Lemongrass (*Cymbopogon citratus*)	Leaf	1 oz.	Steep	10
Lime (*Citrus* spp.)	Zest, fresh	1 oz.	Steep	15
Mint (*Mentha piperita*)	Leaf	.5 oz.	Steep	10
Nutmeg (*Myristica fragrans*)	Pod, ground	.1 oz.	Boil	30
Oak (*Quercus* spp.)	Wood chips or cubes toasted	3 oz.	Secondary fermenter	20–40 days
Rosemary (*Rosmarinus officinalis*)	Leaf	1 oz.	Boil	45
Sage (*Salvia officinalis*)	Leaf	2 oz.	Steep	15
Spruce (*Picea* spp.)	Buds, fresh branch tips extracted in boiling water	6 oz.	Boil	60
Star Anise (*Illicium verum*)	Pod, crushed	1 oz.	Boil	30
Thyme (*Thymus* spp.)	Leaf, fresh	.25 oz.	Steep	10
Vanilla (*Vanilla planifolia*)	Bean, whole extracted in alcohol	2 beans	Secondary fermenter	n/a

TIPS FROM *BYO*

COFFEE AND BEER

When brewing with coffee, the following are the most important things to consider: the style of beer, coffee varietal, roast level, and coffee addition method. Start by considering the style of beer. The obvious choices are stouts and porters. It's no surprise that these meta-styles lend themselves well to working with coffee as highly roasted grains yield deep, rich, roasty flavors and aromas that pair well with the flavors and aromas of coffee. That said, you shouldn't feel constrained with those two choices, as coffee has been used with success in IPAs, Belgian-style beers, and even lagers.

The next option to consider is which coffee to use. If you want a high-quality coffee flavor and aroma in your beer, then you need to use high-quality coffee beans. Opinions on the best varieties are going to span across a broad spectrum, as it's simply a matter of taste, but try to look for single-origin, whole Arabica beans. Avoid flavored varieties, which are often lower quality and may be made with artificial flavoring.

Once you have selected the coffee varietal, you must decide what kind of roast you want. Many brewers tend to favor darker roasts such as French and espresso, because outwardly these provide the boldest flavors. Something to consider, however, is that a light-to-medium-roast bean provides a more complex coffee flavor and aroma because fewer of the aromatic compounds are destroyed during the roasting process. A light-to-medium-roast coffee will also have less burnt or acrid flavors that could give a harsh, bitter character to your beer. If a stronger coffee flavor is desired, then you can simply use more beans.

There are many ways coffee can be added once the beans are chosen. The most common are adding freshly brewed hot coffee directly to the beer at packaging; adding a cold brewed coffee at packaging; or dry beaning, which is adding coffee beans directly to the fermenter as you would when dry hopping.

Coffee meant for a morning cup of joe is generally brewed with hot water, because this quickly dissolves the flavor and aroma compounds (as well as the caffeine) that you are looking for. When making coffee beer, using coffee brewed in this way can certainly provide satisfactory results; however, it's not necessarily always the best way to go as the hot brewing process releases many of those flavors and aromas you want to get in the beer into the air. Cold brewing, on the other hand, is a process of extracting the coffee from the beans with cool (or room temperature) water. This process preserves more of the flavor and aroma compounds and lowers acidity. The third option, dry beaning, involves adding lightly cracked, coarsely ground, or often whole coffee beans directly to your beer, typically in the secondary fermenter. This method yields a complex, long-lived coffee flavor, which may be due to the alcohol in the beer extracting flavor compounds that water alone cannot.

PASTRY BEERS

"Pastry beers" (and they're not all stouts, by the way) are brewed with adjunct culinary ingredients and spices, featuring a flavor profile reminiscent of baked goods or their identifiable ingredients.

Traditional beer ingredients can offer up some of these flavors, but adjuncts are what really separate these beers from other styles. Two of the most common are lactose and vanilla. There's a smooth, soft sweetness to lactose that you just won't get from other sources, and it (or you could go with maltodextrin) will also add body that is important to this style. Vanilla is often used as it creates an impression of sweetness (but isn't actually that sweet, so cloying finishes are less likely). In addition, vanilla is a ubiquitous baking/pastry ingredient, so it's going to do a lot of the "pastry impression" heavy lifting for you. Whatever adjuncts you use, consider the final flavor impact, in terms of both the targeted flavor and secondary/tertiary flavors and textures. Here are some other considerations when designing your pastry beer:

- Start with what traditional beer ingredients can already offer in terms of flavors. We can already count on cocoa, bread, banana, cherry, butter, coffee, caramel, sugar, raisin, and a lot more from traditional brewing ingredients. The advantage of leveraging these first is that we have a dense base of brewing history and experience and knowledge to apply to their use. Start with what you know.
- Plan your adjunct additions ahead of time. Usually these will be added on the cold side after fermentation is complete. They may need to be added in stages as some will require more contact time than others.

- Don't assume these beers have to be high-ABV. Yes, there are a lot of strong, sweet beers out there–many brewers take advantage of the fact that they're not hiding sweetness and leverage the native sweet flavor of ethanol to create a certain impression–but it isn't a requirement. Impressions of sweetness can come from natural process-created flavors (esters), added sugars (lactose), and more.
- In terms of hops, think fruit. Fruit is complementary to nearly all of the typical pastry, chocolate, coffee, and sugar flavors that we are potentially adding as well. Even if you're not using flavor hops and are just adding a few IBUs of balancing bitterness, some small measure of flavor might translate through.
- Use more fruit. This is a surprisingly underfruited sector that leans far more on chocolate, coffee, and caramel than it does on fruit. Fruit adds terrific complementary flavors, especially in lower-ABV or lighter-colored varieties, and so long as you consider the flavor impact of tartness and tannins and the recipe impact of additional fermentable sugars on ABV, you can add a lot of complexity and evocative flavors without too much effort.
- This tip comes from Ben Romano of Angry Chair Brewing in Tampa, Florida, and we think it's a great one. "I like to make whatever dessert I'm trying to emulate in a beer. Understanding every ingredient and how they interact with each other helps me to come up with the adjunct profile that I need and how to accomplish certain flavors."

Oak can take quite a bit of time to infuse its flavor into the beer, but be patient.

WOOD AND SOUR BEER

Beyond traditional ingredients and techniques, you can also manipulate the flavors of your homebrews by oak aging your beer with either a barrel or oak alternative. You can also experiment with non-*Saccharomyces* yeasts to make "sour" beers.

USING OAK

It seems that every year more and more craft brewers are maturing at least one of their beers in unlined wooden casks. A lot of craft brewers especially like aging their beer in used bourbon or other spirits barrels, which adds the spirit's character in addition to wood. In the following paragraphs, we'll take a look at your options for re-creating some of these flavors on the homebrew level.

ADDING OAK FLAVORS

The most direct way to add oak flavor to a beer the same way as a commercial brewery is to age your beer in an oak barrel. However, that isn't always practical for a homebrewer. Small oak casks (5 to 10 gallons/19 to 38 L) are available, but they can be difficult to use. The first problem is surface area; a 5-gallon (19 L) cask has about twice the surface area per unit volume of that of a typical commercial barrel. This means that it is easy to overdo the "oaking" and to overwhelm all other flavor components. Also, because of the surface area effect, evaporative loss of beer through the porous staves of the barrel can be significant during long aging and will also result in loss of carbonation. Finally, the first use of a wooden barrel will take out quite a bit of the extractable oak flavors, so it cannot be used many times for the same purpose (three or four is the maximum, with diminishing results each batch).

A more common approach, which avoids most of these problems and whose effects are easier to control, is to add oak in some form or another directly to the beer. Known as oak alternatives, these options include oak chips, cubes, staves, and spirals, and they are available at most well-stocked homebrew suppliers. These are all much cheaper ($3 to $10, depending upon type) than a barrel, which can cost more than $200. The flavors to be expected from these will vary according to the time of immersion (all are usually added in a secondary fermenter). But the materials themselves, especially the cubes, come in a wide variety depending upon the source (commonly American, French, and Hungarian oaks) and on the degree of toasting (light, medium, and heavy). There are also oak products that have been soaked with whisky and Sherry flavors, or you can soak these items in spirits of your choosing.

As to what type of oak and degree of toast you should try, that is a purely personal choice, decided by your own taste. Try starting with American oak at the lighter end of the toasting range. When it comes to addition rate and residence time in the beer, start at the lower end of the range— you can always work up from there with later brews. As always, make detailed tasting notes on your early efforts in order to guide you on the next beer. Above all, do not assume that more is always better, and remember that the degree of oaking required will depend upon the style of the beer you are brewing. Adding oak flavor is definitely not a procedure where one size fits all.

WHICH BEER SHOULD YOU OAK?

As with most changes in your approach to adding extra flavor to a beer, you should always think carefully about what it is you want to achieve; do not just charge ahead or you may end up with something undrinkable. Mainly you need to think about the normal flavor of the style you are planning to brew and whether oak flavor would throw it out of balance or add some welcome complexity.

The main flavors we are concerned with come from vanillin (that is, vanilla-like) and tannins (which impart astringency), though there are many other less well-defined contributors to oak-derived flavor, such as pepper and roasted notes. Still, if you bear vanilla and astringency in mind, it will be clearer as to which of your beers will benefit from oaking. There are no hard and fast rules, and much will depend upon your own taste threshold and like or dislike for these main flavors.

Low-alcohol beers (below 4.5 percent ABV), such as milds, English brown ales, ordinary bitters, and cream ales, generally suffer from this procedure. They will likely be dominated by oak flavors and are prone to being overwhelmed by too much tannin. Much the same is true for light-flavored lagers, such as American Pilsner and Kölsch, as well as for the various forms of wheat beer (which are meant to showcase other flavors, particularly those derived from the yeast used). Any beer where full-bodied, well-balanced maltiness is the normal characteristics such as bock beers, Scotch ales, and wee heavies, will not really benefit from excessive amounts of vanilla and astringency either.

Hoppy pale beers are another story, however. Even a relatively low-alcohol beer like pale ale can benefit from a little oaking. Go very gently here, though, and opt for light-toasted American oak cubes at a low rate, say 1 ounce (28 g), and let it sit only for 2 to 3 weeks on the cubes. More highly hopped brews, such as IPAs, double and imperial IPAs, and black IPAs, can be oaked to advantage. The high levels of bitterness and hop character will tend to hide astringency from the tannins, and the vanilla and roasted notes from the oak will help to smooth out the hop bitterness.

Oak is risky in fruit beers as it can easily overwhelm the desired fruit character. I know a number of craft brewers have barrel-aged fruit beers, but it is usually done when the brewers are looking for more funky flavors from *Brettanomyces* yeasts, which ferment very slowly. You might be thinking at this point that only big, high-alcohol beers will benefit from oak contact. But that isn't true for Belgian golden and tripel beers, since these are generally designed to be dry tasting (often via a candi sugar addition). That allows the flavors from the Belgian yeasts to come through, and oak tannins would certainly smother those flavors. Of course, high-alcohol beers such as barleywines and imperial stouts can be improved by adding oak flavors, even those that are not highly hopped. That's because such beers tend to be quite sweet and the tannin bite helps make them seem drier, allowing the vanilla flavor to come through nicely.

Above all, beers using highly roasted malts (chocolate, black malts) are the prime candidates for oaking. All the flavors that can be conferred by oak tend to balance out the harshness of the roasted character and add to the depth of the beer's palate. But again you have to be careful about which beer you pick for this. It is easy to spoil a well-balanced, low-alcohol (say, 4.3 percent ABV) brown porter. You would be better off considering a robust porter at the top end of the alcohol range, perhaps around 6 percent ABV. Baltic porters can sometimes be a little bland and many could be improved by judicious oaking. An Irish dry stout probably should not be oaked, since its characteristic flavor is only that of high-roasted malt. Cream stouts, with their typical luscious flavor from unfermentable lactose, are not likely to be improved by adding astringency from the oak. The same might be said for oatmeal stouts since the oatmeal is usually incorporated to give the beer some extra smoothness. However, that does not mean that this style cannot be improved by adding oak as long as you do so at the low end of the addition rate and residence time (say adding one spiral in 5 gallons (19 L) and letting it sit in the beer for no more than 3 to 4 weeks). Finally, a style that really works very well with added oak flavor is imperial stout (or imperial porter). These beers can carry the strong oak flavors very well, even the astringency from the tannins. In fact, since these beers are often aged over long periods, the tannins will degrade and lose their harshness (as happens in aged red wines).

SOUR BEER

In the vast majority of beer styles, the only microorganism that should be present in wort or beer is brewer's yeast, *Saccharomyces cerevisiae*. Increasingly though brewers are looking toward different yeast and bacteria strains for various reasons. One of those reasons is in the production of sour beer where lactic acid bacteria is encouraged to grow. Lambic, oud bruin, Flander's red ale, Gose, and Berliner weisse are all styles of beer that get their sourness primarily from lactic acid bacteria. (The first three also have other "bugs" growing in them that influence the character of the beer.) In this section, we'll go over how to manage the production of sour homebrew, with a focus on the lactic acid bacteria.

UNDERSTANDING LACTIC ACID BACTERIA

There are six principal genra of lactic acid bacteria: *Streptococcus*, *Enterococcus*, *Leuconostoc*, *Lactococcus*, *Pediococcus*, and *Lactobacillus*. Bacteria from the genera *Pediococcus* and *Lactobacillus* are the most common brewery contaminants of the lactic acid bacteria.

All of the lactic acid bacteria grow in the absence of oxygen. (They are anerobic, in the lingo.) Unlike most anaerobes, however, they are not inhibited by oxygen. (They are aerotolerant anaerobes.) As such they can grow whether or not oxygen is present. They don't, however, switch from fermentation to aerobic respiration in the presence of oxygen.

All lactic acid bacteria ferment glucose and produce lactic acid. Some also produce ethanol and CO_2 along with the lactic acid. Bacteria that produce only (or mainly) lactic acid are called *homofermentative*. (The prefix *homo* indicates that they produce only one fermentation product.) Bacteria that produce other products along with lactic acid are called *heterofermentative*. For a brewer, the big difference is that heterofermentative "bugs" produce carbon dioxide (CO_2). If you've ever had a sour, gushing beer, it's fairly likely you had one of these bacteria as the contaminant.

Bacteria from the *Pediococcus* are homofermentative. The genus *Lactobacillus* is made up of mostly homofermentative species, although there is a subgroup of heterofermentative species. Beer contaminated with any of these bacteria will turn sour from the production of lactic acid. Beer contaminated with heterofermentative bacteria will also have excess dissolved CO_2. In addition, lactic acid bacteria—especially *Pediococcus* and *Leuconostoc*—may also produce other products such as diacetyl.

The genus *Lactobacillus* is divided into three major subgroups. The species of *Lactobacillus* most relevant to brewing belongs to a group in which lactic acid is produced from over 85 percent of the glucose fermented. This group—including the species *L. delbrückii* and *L. acidophilus*—grows well at higher temperatures (90 to 120°F/32 to 49°C), but much more slowly (if at all) at cool temperatures (below 60°F/16°C). *L. delbrückii* is a common brewery contaminant and a microorganism used in the production of Berliner weisses.

TRADITIONALLY SOURED BEERS

It's ironic that although it's impossible to completely eliminate lactic acid bacteria from a brewery, they can be equally difficult to grow when you really want them to. At best, sour beers take time to develop. At worst, a suitable level of sourness never develops. This is because lactic acid bacteria such as *Pediococcus* and *Lactobacillus* have strict growth requirements. They require amino acids and trace vitamins, and both of these are scarce in a finished beer. Homebrewers get around this limitation in a few ways when making a sour beer. The first, and most obvious, solution is to always pitch enough suitable microbes.

The second solution is to let the beer sit on its yeast sediment for long periods of time. As yeast cells die, they release nutrients that get scavenged by the lactic acid bacteria. After 6 months on the yeast, most sour beers will have a substantial amount of sourness. After a year, they will likely be sour enough for all but the biggest sour fiends. A benefit of this method is that it is simple—just leave your beer in primary. A drawback is that dead yeast can add an undesirable nutty flavor to beer. Another drawback to this is that the amount of time spent on the yeast isn't a reliable way to control the acidity of a beer.

Some of the strains of lactic acid bacteria that grow in wort or beer, most notably *Lactobacillus*, grow best at high temperatures (90 to 120°F/32 to 49°C). Let your sour beer condition at around 65 to 70°F (18 to 21°C) and try not to let it push past 80°F (27°C) if you are aging for long periods of time if you can help it.

Lactic acid bacteria are, of course, more resistant to acidic conditions than most bacteria. *Lactobacilli* are, in turn, more resistant to acidic conditions than most other lactic acid bacteria. Most are able to grow well at pH values in the 4 to 5 range and grow more weakly at lower pHs. The pH of lambics varies from 3.3 to 3.9, whereas the pH of most "regular" beers varies from 4.0 to 4.4.

Note: *When lactic acid bacteria is encountered in a brewery, it is most often an unwanted contaminant. Lactic acid bacteria, especially those from the species* Pediococcus damnosus, *are common beer spoilers. They are often found in lager fermentations and can cause elevated diacetyl levels. Another species of lactic acid bacteria common to breweries is* Lactobacillus delbrückii. *In addition to producing sourness, beers contaminated with lactic acid bacteria may also go turbid.*

SOURCES OF LACTIC ACID BACTERIA

There are various sources of *Lactobacillus* and different methods to produce a sour beer. You can get a supply of lactic acid bacteria from a laboratory, a bottle culture, from nature, yogurt, probiotics, or it can even be found on unmashed grains (it's found on malt husks). There are now even yeast strains, naturally occurring and bioengineered, available that produce alcohol and lactic acid at the same time during fermentation. However, sour beer production can often be broken down into two separate categories: (1) traditional methods and (2) quick souring methods. The traditional methods are widely known to be co-fermentation from spontaneous or cultured sources often with barrel/ foeder aging. These methods typically have mixed cultures of *Saccharomyces*, lactic acid bacteria, and even *Brettanomyces* and *Pediococcus*. A positive of this wide spectrum of microflora is that it tends to produce a more complex, flavorful beer. However, it takes much longer, and some brewers don't have that extra time or space to produce these. The other concern is consistency. It's not always going to be the same as brews before. This is where the popular "quick souring" methods come into play. Typically, this is done by a mash- or kettle-souring process. Kettle souring has become one of the more popular methods due to the quick turnaround times and typically consistent fermentations. However, these methods have also been noted to be a bit one-dimensional in aroma and taste in comparison with the flavor profiles of a co-fermented or barrel-aged sour beer (find more on kettle souring below).

KETTLE SOURING

Kettle (or brewhouse) souring is usually done in the brew kettle. The mash is completed as one would in normal brewhouse operations. The wort is then separated from the grain, run into the kettle, and cooled to the proper temperature (generally between 95–120°F/35–49°C, depending on the strain). Then, a source of *Lactobacillus* bacteria is pitched into the wort. The wort is held at that temperature and the *Lactobacillus* is allowed to produce lactic acid until the desired pH is achieved. At that point the wort can be boiled like a normal brew. The bacteria are killed by the heat of boiling, and the wort can then be transferred into a fermenter.

The advantages of kettle souring, as opposed to post-kettle souring, are several. First, the soured wort gets boiled before being transferred to fermentation, thus no live *Lactobacillus* bacteria is sent into the fermenter. Second, it enables you to fix the level of acidity desired with a high degree of accuracy. Third, if desired you can add a little more hop bitterness to your beer. This can be done during the boil because the usually hop-intolerant *Lactobacillus* has already done the souring prior to your hop additions. It is important to know that as little as 5 IBUs can retard *Lactobacillus* growth and that sour beers are traditionally not hoppy beers.

As far as quick souring methods go, the reason kettle souring of the wort is superior to sour mashing (in which *Lactobacillus* is introduced to the mash after the conversion of starch to sugars and dextrins is complete) is that the mash contains a lot of bacteria that can produce some pretty unpleasant aromas in beer. In order for these bacteria to produce these notes they need two things: a high pH (above pH 5) and oxygen. Without these two vital things these "bad" bacteria cannot carry out their life functions and produce unwanted aromas. To eliminate these two things a brewer needs to get the wort off of the grain (thereby greatly reducing the amount of oxygen and bad bacteria in contact with the wort) and begin the souring process as fast as possible so as to lower the pH below 5.

Removing the wort from an oxygen source is probably the more important of the two. The kettle dimensions provide less surface area-to-air (oxygen) ratio. Some brewers go a step further and either purge their kettle with a gas that will displace the air (like carbon dioxide, nitrogen, or argon) or put a "blanket" of those gases on top of the wort as the kettle fills and during the souring. Some brewers even do both. Getting the wort into the kettle, purging it with gas, and/or blanketing the wort with gas will all help to exclude oxygen from your process. Homebrewers have the advantage of the versatile corny kegs available to them where oxygen can be purged from the vessel and sealed for this purpose. Find more on kettle souring in a corny keg at https://byo.com/article/the-lacto-lounge/.

There are several options to lowering the pH below 5 as soon as possible. The easiest is to add some food-grade lactic or phosphoric acid to the kettle as you run off your wort—just enough to adjust the pH down to 4.9. The other way to lower the pH quickly is to pitch the proper amount of LAB into the wort so it begins souring it fast. In either case you will need to lower the temperature of the wort from around 150°F (65.5°C) to the optimal range for LAB growth depending on the strain. This can be done in several ways: You can create a wort with a higher gravity than is ultimately desired and add cooler water to the wort in the kettle to cool it down. Or you could run the wort through a heat exchanger to cool it to the proper temperature. What you should not do is wait for it to cool on its own.

CONTAMINANTS AND DECONTAMINATION

The low pH of sour beers renders them uninhabitable to most stray contaminants. However, aging sour beers are prone to surface contamination by *Acetobacter* and other aerobic bacteria and yeasts. *Acetobacter* can live only on the surface of beer, where it has access to oxygen. As its name implies, it produces acetic acid, which can yield a vinegar flavor to a beer. Because traditionally soured beers take a long time to age, occasionally the fermentation lock will dry out and a bloom of *Acetobacter* will appear. Likewise, many times sour beers will be opened for "feeding" or to add fruits, and this presents an opportunity for contamination.

If you get *Acetobacter* or other surface contamination, don't panic. A small amount of acetic acid in a sour beer is not going to wreck it. If you were planning to bottle the beer soon, just leave it alone. If you're using an airlock, be sure to keep it full. When aging in a carboy, make sure the fermenter is "topped up"; in other words, the beer level if found above the curvature in the neck of the carboy. This will minimize any potential oxygen exposure. If aging in an oak barrel, be sure to "top up" the barrel with a similar beer every month or so to minimize the beer surface area exposed to air. If you find that *Acetobacter* has bloomed on the surface of your beer, you can minimize the impact by carefully racking out from under the surface contamination and into another container. Leave the last little bit of beer behind. Minimizing the beer's subsequent exposure to oxygen will also help you keep surface blooms in check, so be sure to keep your souring beer fermenters topped up.

Many homebrewers worry that if they make a sour beer, all their subsequent beers will turn sour. Good cleaning and sanitizing is enough to decontaminate glass, stainless steel, and plastic that is in good shape. On the other hand, a barrel that has gone sour cannot be brought back due to how porous the wood is, so it should remain in use for only sour beers. If you are worried, though, you can mark and isolate any porous items (stoppers, scratched buckets) and use them only for sour beers. Finally, for a little piece of mind, you can treat your next nonsour beer with a dose of lysozyme. Lysozyme will kill all gram-positive bacteria, including lactic acid bacteria. A small preventive dose, just 1 teaspoon per 5 gallons (19 L), added to the fermenter will do the job.

OTHER SOURCES OF SOUR

Whenever people are faced with a long process in which risks are involved, it's common for them to ask, "Can't I just cheat?" In the case of sour beers, yes, you can (sort of). If you want to skip the long aging times—and not actively grow lactic acid bacteria in your brewery—you can simply add lactic acid to your beer. In a 5-gallon (19 L) batch, it only takes a few ounces (about 60 mL) of an 88 percent lactic acid solution to get a noticeable tang to the beer and a few more drops to get a pronounced sourness.

Lactic acid is the same whether you get it from growing lactic acid bacteria in your wort or from a bottle. However, a bacteria-soured beer will have a dry mouthfeel as the bacteria will have consumed some or all of residual carbohydrates in the beer. A beer soured by adding acid will have a fuller body and likely have a bit sweeter edge to it. This is not necessarily a bad thing; sweet and sour are two flavors that go well together. However, you cannot closely reproduce any classic sour beer styles simply by adding lactic acid to a regular beer.

PACKAGING

Two fairly common problems occur when bottling sour beers: lack of carbonation or overcarbonation. Because the pH of sour beers gets low, and because they condition for a while, there may not be enough healthy yeast to carbonate the beer after packaging. Adding a teaspoon of fresh yeast per 5 gallons (19 L) of beer at bottling can help. Any dried ale yeast will work well.

Conversely, because some CO_2-producing lactic acid bacteria can continue to work on residual carbohydrates in a beer, sour beers can sometimes overcarbonate. This is usually only a problem if the beer is bottled too soon. Before bottling, taste a sample of the beer. If it does not taste very sour and still has nearly as much body as you would expect without the bacteria added, don't bottle it.

TIPS FROM *BYO*

SMOOTHIE SOURS

In the early 2010s, breweries resurrected the nearly forgotten German beer styles of Berliner Weisse and Gose. As these "kettle soured" beers grew in popularity, brewers began expanding the style with additions of fruit of all different types. As the style continued to grow, brewers began pushing the limit of how fruit could be perceived in sour ales. Eventually the level of fruit being used led to a near smoothie-like character, inevitably being called smoothie sour beer. From there the use of crazy adjuncts and ingredients such as cinnamon, chocolate, ice cream, and cheesecake (just to scratch the surface on the options) no longer stood in the way.

SOURING THE BASE BEER

Smoothie sours differ from the traditional sour beer base in the fact that you are trying to create a beer that is well attenuated yet has a full mouthfeel. The base recipe of these styles is similar to that of a hazy IPA recipe, chock full of wheat and oats: a little bit of acid malt for mash pH adjustment and some dextrin malt for the tremendous foam retention that you will get from the fruit additions. The base beer is commonly 9.75–10.5 percent ABV, with the unfermented fruit addition added later accounting for about 45–50 percent of the final volume of finished beer, creating a beer that is about 5 percent ABV by the time it is packaged. You are also looking for wort that has a bunch of fermentable sugar. Shoot for a low mash temperature in the 148–149°F (64–65°C) range. This will help the yeast beat the acidity and high gravity at the same time.

After a very short boil (40 minutes or so), send the wort to a souring vessel and add your favorite *Lactobacillus* culture for your sour beer. A good target pH to aim for is from 3.40 to 3.55. Any lower than 3.40 and you end up with a base beer that does not interact well with the acidity of nearly any fruit; any higher than 3.55 and you start to lose the essence of the base sour beer. If you do not have a pH meter, go by taste.

After you have reached your optimal souring point (depending on the culture this could be in 18–48 hours), bring the wort back to your kettle and begin to boil. A short 40-minute boil works great for these beers. The kettle additions are simple: dextrose (corn sugar), yeast nutrient, and whirlfloc. You want to incorporate dextrose as approximately 9–10 percent of your total fermentable sugars. The goal is to have a beer out of the kettle at 1.097–1.100 specific gravity. After fermentation you'll have a 10-ish percent ABV beer.

FERMENTATION AND THE ROAD TO ADJUNCTS

Make sure that your wort is thoroughly aerated with oxygen. The finished beer will need it, as the fermentation will be tough on the yeast with such a high original gravity and low pH. The target final gravity is between 1.020 and 1.024 SG. A clean yeast such as the "Chico strain" will work well. It can absolutely withstand the high gravity and low pH and achieve the desired fermentation metrics.

Once you have reached terminal gravity, it is time to start thinking about your fruit and other adjunct additions. To really create a smoothie sour beer, your fruit is not meant to be fermented. It needs to stay in the full purée state to really get the smoothie feel to encapsulate the drinking experience. If you are simply going to add fruit, the recommendation would be to crash the beer down to 34°F (1°C) and transfer off the yeast cake to secondary conditioning to allow more yeast to settle. After another 4–5 days cold, rack the beer off the yeast cake to fruit.

Starting out with your first smoothie sours, keep it simple when it comes to adjuncts and start by using only fruit. Using 100 percent raspberries or blueberries can taste fantastic. If you feel a tad more adventurous, go with known fruit combinations like POG (passion fruit, orange, guava).

The selection of how the fruit is prepared is important. Fresh fruit is not advisable for smoothie sours as there may be wild yeast on the skins that will cause refermentation. The ideal fruit is aseptic purée, which can be found in most grocery stores.

When you have the fruit additions down and are ready to be more adventurous with your adjunct usage, there are a few extra steps that are necessary to really bring out the full flavor of all the adjuncts. After fermentation is done and you see a steady FG for 3 days straight, let the beer hang on for another 4 days at fermentation temperature. Do not cool or crash the beer. This is going to be integral to the pickup of flavor from your secondary conditioning. As you have carefully selected your adjunct additions, place them in your secondary vessel, and if you have the ability, then purge the vessel with CO_2. Transfer your base beer on to the adjuncts for 5–7 days, which will allow for the complete absorption of the flavors without those characteristics becoming overbearing on the base beer. There are certain adjuncts that can be detrimental to your beer out of the gate if left to too much time. Mainly be careful with spices such as cinnamon, nutmeg, and allspice. After your 5–7 days you are ready to follow the previously mentioned steps and head to fruit.

PACKAGING

Once your newly created smoothie sour is near ready to drink, it's time to consider how it will be conditioned and poured. The best choice for homebrewers is to condition in a corny keg. Add the fruit purée to the keg and rack the beer on top of it. Once racked and in your kegerator you can begin conditioning. The main point to consider is separation. During conditioning and pouring you will want to make sure you are always shaking the kegs to keep everything mixed. Carbonation level is based on personal preference, bubbles or no bubbles, but be sure to keep the keg cold at all times after kegging.

Bottling these smoothie beers is a completely different story. There will be no bottle conditioning as the yeast will rip through the fruit and cause literal bottle bombs (really, please don't attempt to bottle this way). If you want bottles, counterpressure fill the bottles from the keg.

RECIPES

HOMEBREW RECIPE DESIGN

We all know you make beer from barley, hops, yeast, and water, but some of us don't know where to start when designing a beer on our own.

THE TWO SCHOOLS OF BEER RECIPE DESIGN

In the broad sense, there are two basic approaches you can take to beer recipe design. Let's call these the artistic and the mechanical approaches.

The *artistic approach* involves pushing the boundaries of a beer—that is, using new or unusual ingredients in outrageous ways to create something new and unique. This is where many homebrewers excel since they are willing to take risks, experiment, and try unconventional things. Jalapeños in your beer? Why not? Maybe it will go well with pizza! Try juice, berries, cinnamon, fennel, chili, chamomile, or any other flavor combination you can think of. Brewers in this camp often make beer from whatever they have around, throwing ingredients together in rough proportions on the spur of the moment. This is how an artist might approach making beer.

In contrast, when utilizing the *mechanical approach*, a brewer must be much more methodical in the beer's design. These recipe designers start with a known beer style or commercial beer they want to make, carefully research the ingredients and proportions used, and then calculate the bitterness, predicted gravities, and color to get exactly the combination they desire. It's an engineer's approach to beer making.

The artistic school is perhaps embodied best in Randy Mosher's book *Radical Brewing*, while the mechanical approach is best captured by Ray Daniels's book *Designing Great Beers*. This isn't to say one approach is better or worse—honestly, saying that there are only two paths to follow is an overly simplistic way of looking at the situation, because designing a recipe often involves both schools of thought. No one would say that John Coltrane was not artistic because he understood and applied music theory to his improvisational style! Likewise, *Radical Brewing* does not require a brewer to be ignorant of brewing calculations—those calculations can, in fact, be applied to funky ingredients like peppers to help maintain balance.

Many of the best brewers do achieve a balance between art and science, much like Coltrane did with music; they know the numbers, but they also know how to use the ingredients to create something that is more than the sum of its parts.

A STRUCTURED APPROACH TO BEER RECIPE DESIGN

For those of you just starting out in beer recipe design, it's important to understand a structured (i.e., more mechanical) approach to building recipes. There will be time to add more artistry, complex ingredients, and techniques as you become more advanced and build up your understanding of ingredients. However, taking a structured approach will make it easier for a less-experienced brewer to build a new recipe.

The basic process to follow when building a new homebrew recipe is a series of six steps:

1. Start with a well-defined goal for the beer.
2. Research the target style and beer.
3. Select the ingredients.
4. Develop the specific grain bill, hop schedule, and fermentation schedule.
5. Apply specific techniques to help enhance the beer.
6. Brew, judge the beer, and iterate.

Let's look at each of those steps in depth.

START WITH A DEFINED GOAL

The first and most important step in creating great beer is to start with a well-defined goal. If you are stumped over what to brew next, do some personal reconnaissance at your local supermarket, craft beer bar, or homebrewing event to sample a variety of beers and find something that piques your interest. Some examples of brewing goals include:

- Making a clone of a popular commercial beer
- Brewing a specific style
- Making a beer for a homebrew competition
- Designing a beer centered around a particular ingredient (such as smoked oats)
- Creating a beer to serve at a certain event (like an Oktoberfest)
- Inventing something really unique (Jalapeño-Flavored Atomic Hop Bomb)

Write down a line or two defining what you are trying to do before you start coming up with a recipe. It could be really simple, like "Bass Ale clone," or it could include a detailed description of the style in the case of a competitive beer. Define what is different or unique about your beer so that you have a solid plan to stick to when you start choosing ingredients.

For many brewers, designing new recipes is as much fun as brewing!

RESEARCH THE TARGET STYLE OR BEER

Once you have a clear idea or goal of what you want to brew, research the beer you're going to make. A good first stop is the style guide from the Beer Judge Certification Program (BJCP) at www.bjcp.org. Each beer description includes aroma, appearance, flavor, history, and key ingredients as well as vital statistics such as bitterness level, original and final gravity, and color ranges for each beer style.

Books are another major resource, including design books such as those mentioned earlier (Ray Daniels's *Designing Great Beers*) and style and recipe books such as *Brewing Classic Styles*, by Jamil Zainasheff and John Palmer, Jamil's *Best of Brew Your Own* special issue, *30 Great Beer Styles*, or even the recipes found starting on page 156 of this book. If you like a particular beer style, you can also consider the many style-specific books such as *IPA: Brewing Techniques, Recipes and the Evolution of India Pale Ale*, by Mitch Steele, or any of the Classic Beer Style series of books.

If you don't have a large brewing library, you can also find a huge amount of brewing information online. A quick search will reveal articles on brewing various beer styles, blogs, forums, and recipe sites, such as BeerSmithRecipes.com and—of course!—www.byo.com. Compile a few recipes that are similar to the beer you are trying to brew, then take a close look at which ingredients they used and in what proportions.

Finally, the tastiest method of research is firsthand research. Sample some beers that are similar to yours from supermarkets, local craft breweries, or other homebrewers. With practice, you will be able to discern what the major ingredients are and get a good feel for what the beer should taste like. This knowledge will come in handy when you go back later to judge and improve your own recipe.

SELECT POTENTIAL INGREDIENTS

At this point, make a complete list of the ingredients you want to use in the recipe, including the yeast. (More advanced brewers can also include the water profile. More on this on page 124.) Don't try to determine the exact proportions yet—just focus on identifying the key ingredients and what they contribute.

Start with the ingredients that define the style. For example, you can't make a Belgian wit without unmalted wheat or a Bavarian weizen without the proper yeast to provide the banana/clove flavor. A complex, fruity ale yeast is a key ingredient for a classic English ale. This is where your research will really come in handy.

Next, consider what alternatives might be possible. Could you substitute an American hop variety for an English one or use a specific barley variety such as Maris Otter to give the beer more character? Could you add something really unique and off-the-wall like peppers or cocoa?

Finally, go through the notes and ideas that you've come up with and simplify. Many beginning brewers try to put everything but the kitchen sink into their beer. In contrast, most commercial beers are made with just a handful of ingredients, partly because commercial brewers can't afford to maintain a huge inventory of dozens of types of grains and hops but also because simpler is generally better. The best homebrewers are also ruthless in simplifying their recipes: they don't add any ingredients to the beer without a specific purpose in mind.

One way to get used to brewing with fewer ingredients as well as to understand your ingredients is to explore the Single Malt and Single Hops (SMaSH) style of brewing. With SMaSH style, you brew a beer with only one malt and one hop variety. A surprisingly wide variety of styles can be made, including Pilsner (all kinds), Vienna lager, saison, Munich dunkel, wild ales, IPAs, and even barleywine. The advantage of SMaSH is that it gives you a really great feel for what each base malt and hop adds to the overall beer. It also gives you a good appreciation of what can be accomplished with few ingredients.

DEVELOP THE SPECIFIC GRAIN BILL, HOP SCHEDULE, AND PROPORTIONS

After simplifying the ingredient list, the next step is to develop a specific recipe. At this point, it is recommended to use some kind of recipe software or spreadsheet. No matter what package or spreadsheet you decide to use, some kind of system for estimating the bitterness, color, gravity, and alcohol content of your recipe as you build it is critical to designing a good, balanced recipe. It is even better if your software or spreadsheet can also take into account your specific equipment since equipment volumes can significantly affect these estimates.

Start first with your grain bill (or extracts). It is easiest at this point to start with percentages, such as 80 to 90 percent base grains, and the remaining specialty grains (for example). Once you have the percentages right, you can scale the weights for each grain up to reach your desired original gravity for the style.

Next, work on the hop schedule. Try starting out with a simple hop schedule, for example, two boil additions followed by an aroma addition at flameout. The boil additions provide bitterness, and the aroma addition provides the fragile hop aromas.

Recent research indicates that most of the hop aroma oils you want to preserve with late hop additions are boiled off in just a few minutes. This means that rather than adding hops during the last 20, 15, 10, or 5 minutes of the boil, you're better off steeping hops at flameout or using whirlpool additions. These options preserve the critical aroma oils in the beer. (Read more about these techniques starting on page 88.)

Similarly, long exposure periods are a thing of the past with dry hopping. Recent research indicates that shorter dry hop contact times result in a better overall aroma and flavor. In fact, as little as 24 hours of contact time when dry hopping can deliver hop aroma. (You can also read more about dry hopping starting on page 92.)

When selecting a yeast, you can rarely go wrong selecting the traditional strain to match your target style. However, sometimes it is worthwhile to go "off-style" to achieve a particular effect. For example, I might use a dry, clean yeast like a California ale yeast if I want a very clean finish—even if I'm not brewing a California ale.

A yeast starter is very important when working with liquid yeast. Liquid yeast has a shelf life of only 6 months and degrades in viability about 20 percent every month. A typical homebrew yeast package has about 100 billion cells when produced, but that can degrade to 50 billion or so by the end of its life. In contrast, the ideal pitch rate for a 1.048 American ale is about 164 billion cells, so a 1- to 2L starter is required if you want to pitch the ideal amount of yeast. Dry yeast does not usually require a starter—just hydrate it 15 to 20 minutes before use. (For more about making a yeast starter, see page 42.)

Water is often overlooked as a major beer ingredient; it can be a pretty complicated subject for new and intermediate brewers. Understanding how hard or soft your local water source is and what minerals it contains is critical to making good beer. Hard water is high in calcium (Ca) and magnesium (Mg), which is usually good for brewing, while soft water is low in Ca and Mg, which means the water usually needs to be improved. The actual problem is alkalinity (carbonates): high alkalinity is nearly always bad and low alkalinity (less than 50 ppm) is nearly always good. In general, pale lager beer styles such as Munich helles benefit from lower amounts of minerals in the water, while the typically more assertive ale styles such as IPA and porter benefit from higher levels. If you have water issues, consider using bottled water. Adjusting water and mash chemistry is a pretty complex topic, but if you would like to learn more, there is a book available: *Water: A Comprehensive Guide for Brewers*, by John Palmer and Colin Kaminski.

APPLY BREWING TECHNIQUES

The last major stage in building a great beer recipe is to decide which techniques best apply to the beer you are brewing. This can be highly subjective because different techniques drive different effects, and you need to know which ones work best with various styles. Again, this is where your background research will come in handy.

For extract brewers, try adding freshly steeped specialty grains to enhance your recipe. Steep grains like caramel, crystal, or chocolate malt, then add the extract and start the boil. Another popular technique is to add the bulk of your extract late in the boil (the last 10 to 15 minutes). This technique is called a late extract addition, and it will reduce the darkening of the wort and the production of atypical flavors that are associated with high-gravity boils intended for later dilution in the fermenter.

All-grain brewers should vary the main mash step temperature to adjust the body of the beer. For a light-body beer, mash in at 148°F (64°C); for a medium-body beer, use 152°F (67°C); for a full-body beer, use 156°F (69°C). Changing the temperature alters the fermentability of the wort, resulting in a light- or full-bodied finished beer.

We recommend a single-step infusion mash for virtually any all-grain beer. If you want to use unmalted grains or cereals with a single-step mash, just use the flaked or torrified versions—these do not require any cereal mashing steps. If you are just getting into all-grain beer, try brew-in-a-bag (BIAB) techniques. Brew-in-a-bag uses a grain bag in the boil pot instead of a separate mash tun. This lets you get into all-grain brewing with less equipment and also saves some time when brewing (see page 37).

For very dark beers, segregating dark grains and steeping them in a "tea" is another great technique to consider. Mashing very dark grains for a long period can lead to additional bitterness, but if you separate them into a steeped tea as described above for extracts, you can reduce the bitterness and get a smoother flavor. This can be a plus for certain dark beers, such as mild porters and milk or sweet stouts.

Depending on the desired style, a variety of hopping techniques are available. I've already mentioned the use of steeped/whirlpool hops after the boil as well as dry hop additions to enhance fresh hop aroma, but obviously, you can also vary boil times to change the bitterness of your beer. The other hop technique I recommend when all-grain brewing is first wort hop (FWH) additions. A FWH addition is added to the boil pot at the beginning of the sparging process, allowing the hop to steep for a short period before boiling. The net effect of FWH is to provide smoother, more pleasing bitterness, making FWH appropriate for beer styles where you are not looking for dominating bitterness. For more about FWH, see page 87.

As far as fermentation goes, we've already mentioned the importance of a good yeast starter.

Aerating your wort with air or oxygen before pitching the yeast will also enhance your fermentation. Temperature control is very important! Ferment in a controlled vessel if possible.

JUDGING AND IMPROVING YOUR RECIPES

It is no coincidence that just about every top-level competitive homebrewer is also a certified beer judge. Judging beer is a separate skill, but if you really want to make great beer, it is one you need to master.

The BJCP score sheet (available at www.bjcp.org) provides great guidelines for evaluating your own homebrews. Use the following evaluation process for tasting your homebrews:

1. Evaluate the external appearance of the beer.
2. Capture the aroma up front, right after the beer is poured.
3. Evaluate color, clarity, and head retention before testing.
4. Taste the beer, noting the overall impression first.
5. Evaluate finish, malt, hops, aroma, mouthfeel, and obvious flaws.

During all of this, take notes on a score sheet so you can honestly evaluate the strengths and weaknesses of your recipe. Use these notes to make adjustments on your next iteration. A lot of homebrewers never brew the same recipe twice, but commercial (and competitive) brewers spend a lot of time perfecting a recipe on a pilot system before they brew 30 or more barrels of it. If you want to brew your best beer, it's wise to follow their lead and brew your recipe multiple times until you get it to where you want it to be. Your goal in judging and iterating is to make a better beer each time you brew a particular recipe.

CLONING RECIPES

It's no secret that homebrewers drink a lot of beer, and not all of it is homebrew. Given that homebrewers also enjoy many craft beers and imports, it's not surprising that a popular topic among us is clone brew recipes—namely, creating homebrew recipes for commercial beers. Recipes for clones of hundreds of commercial beers are available. On the other hand, there are tens of thousands of commercial beers available. So what do you do if nobody has drawn up a clone of your favorite beer? Do it yourself, of course!

WHAT YOU'LL NEED

To formulate a clone recipe, you'll want to use some sort of beer recipe calculator. This can be a standalone program (such as ProMash, BeerSmith, or Strangebrew), or an online calculator (such as www.beertools.com or the Recipator at www.brewery.org). If you can calculate original gravity (OG) and color (in SRM) from the amount of malts in the recipe, final gravity (FG) from the attenuation of the yeast, bittering (in IBUs) from the hops added, and alcohol (in ABV) from the drop in specific gravity, you'll be on your way.

The second and most important thing you'll need to formulate a clone recipe is information . . . and lots of it. To draw up a decent clone recipe, you'll need the aforementioned beer specifications plus information on both the ingredients in the beer and the procedures used to make it. For ingredients, you'll need to know the types and percentage of malts used, the types of hops used and when they are added, the kind of yeast used, and information on any other ingredients (kettle adjuncts, spices, fruits, and so on). On the procedural side, you should find out the details of the mash program, boil times, fermentation temperatures, and any unusual processes used.

WHERE TO GET THE INFORMATION

Information on a commercial beer can come from a variety of sources. First and foremost, you may be able to get much or all of the information straight from the brewer. If your local brewpub has a porter you just love, stop by during the day—or whenever the brewer is most likely to be there—and ask if you can talk to him. Some brewers are reluctant to give out any information about their beers or are even bound by confidentiality agreements not to, but many others are happy to talk shop. Information about a beer may also appear on a brewery's website or on their packaging.

If you can't get any information from the brewer or brewery, you may be able to find at least some information (such as alcohol content, in ABV) on other websites. Recipes for similar beers can help you develop a clone recipe as well. Once you've gathered—or guessed at—all the information you need, you're ready to draw up the first draft of your clone recipe.

IN THE CLONING LAB

Once you have all the information assembled, one way to construct a clone is to use a trial-and-error method of entering ingredient amounts into your brewing calculator until you get the calculated beer statistics right. As an example of how to do this, we'll walk through cloning a real-world beer, Summit Winter Ale, using information found on their website along with tasting notes of this beer.

STRENGTH

Start the construction of your clone by getting a rough idea of how much malt you'll need to reach the beer's original gravity. To do this, enter various amounts of base malt until you get a number equal to your OG estimate. We found out that we would need 12 pounds (5.4 kg) of 2-row pale malt to hit the website-specified OG of 1.058. Extract brewers will use light malt extract as their base malt, but the rest of the process is the same for both extract and all-grain brewers.

What if, however, your only indication of the size of the beer was the alcohol content? Using only the ABV, you can still estimate the starting gravity of a beer. If you know the beer's yeast strain (or can make a reasonable guess), you can use its average attenuation to estimate the OG. In your recipe calculator, select the proper yeast type or type in a reasonable number for attenuation. Then fill in amounts of base malt until you reach the correct ABV.

Next, if you know the percentages of the other malts, multiply the total amount of grain by these percentages and fill them in. For example, if we knew—which we don't—that Summit used 10 percent caramel malt, we'd know to add about 1.1 pounds (0.49 kg) of caramel. If you don't have any information on the proportion of the various malts, the color depth of the beer can help you make a reasonable guess.

COLOR AND MALT FLAVOR

If you don't know the percentage of other malts, start adding the other malts in reasonable amounts into your brewing calculator. (You can use information on how similar beers are brewed as a basis for what a reasonable amount is.) In the case of a beer with one base malt and one specialty malt (both of known color rating), there is only one combination that will yield the right color and gravity for the beer. If there are three or more malts, there are an infinite number of solutions to the puzzle.

Using trial and error, we found that 1.25 pounds (0.57 kg) of crystal malt (75°L) and 1 ounce (28 g) of black patent malt will achieve the right color for Summit Winter Ale. How did we decide to use crystal 75°L? Well, from experience, we knew the amount of black patent malt that would give a nice amount of color but only a tinge of flavor (as noted from tasting Summit Winter). From there, we found by trial and error that when we used crystal 75°L, we got a reasonable amount of crystal in the recipe for a beer of this type.

Note that as you add other malts into your calculations, you will need to decrease the amount of base malt to keep the beer at the correct OG. This can in turn change the color of your beer. With multiple malts, this can lead to a lot of fiddling. However, after you've done this a few times, you'll get better at it.

Once your OG and color match your initial estimates, take a look at the final gravity (FG) and alcohol content (in ABV). If you're lucky, they might be spot-on. If they're not, adjust the amount of attenuation so the FG or ABV is right. In our case, I needed to lower the FG to 1.012 to get the ABV specified on the website.

Sometimes you may enter all of your information into your brewing calculator and the results still won't match up with the brewery's information. For example, you may enter the percentages and color ratings of the malts they use and end up with a calculated color other than the SRM they claim. Likewise, the alcohol content they claim may not jibe with the drop in specific gravity. From the standpoint of cloning, yes, you need to decide how to deal with this discrepancy. However, try forgetting about the numbers for a second and instead focus on formulating a reasonable recipe that will work in your own homebrewery.

BITTERNESS AND HOP CHARACTER

Once you have the malt information set, you can begin to calculate the hopping schedule. Because wort density affects hop utilization, you need to get at least the original gravity of your beer set before you calculate hop additions.

Type different amounts of hops into the brewing calculator until you hit the target IBU. (If you're lucky, the brewer will have specified the amount of IBUs for each addition of hops.) If you don't have any information about the timing of the additions, use information from similar beers as a guide. For example, in the case of the Summit Winter clone, we found that we needed 4.5 AAU of bittering hops (boiled for 60 minutes) combined with 3.75 AAU of flavor hops (boiled for 15 minutes) to reach the 20 IBU specified. Of the hops they had listed, we guessed that the Willamette hops would be the bittering hops and the more flavorful Fuggles and Tettnanger together would be the flavor hops.

Once the malt and hops have been chosen, all that's left is the yeast, water, and perhaps the miscellaneous ingredients—the details of which you either have in your possession or not. Add those details to your recipe. You now have the first draft of your clone recipe!

ASSESSING THE CLONE

Once you've got your clone recipe drawn up, you'll probably wonder how it tastes. The obvious solution is to brew it and find out. However, if you've had to make several assumptions along the way, you may be hesitant to actually brew your recipe, afraid that you will be wasting your time. As you draw up a clone, it's natural to think about all the uncertainties. Once you're done calculating the clone, however, it's good to step back and also think about everything you know you got right before you examine the consequences of being wrong. Let's use our presumptive Summit Winter Ale clone as an example.

We know the OG, ABV, and color of our clone match from the website information. We also know that the malt and hop types are correct. However, we did guess at the color of the crystal malt, and the amount of crystal malt I used was based on that guess. What if I was wrong about that? Different colors of crystal malt are roasted differently and have different flavors, but the different flavors lie along a continuum. Given the color of the beer, the lightest versions of crystal malt can be ruled out. Likewise, the darkest are not that likely seeing as only a small addition of them would be required. So if we guessed wrong, the crystal malt flavor might be different than the actual beer, but not so entirely different that the beer would taste completely off.

Likewise, we guessed at the relative contributions and timing of hops. However, this beer doesn't have a ton of hop bitterness or flavor. It's balanced more toward the malty side (as many winter beers are), so unless we were ludicrously way off on the hops, this shouldn't produce a huge difference in the flavor of the beer. The biggest factor in being off of the target beer lies in the choice of yeast.

Another way of assessing your clone before brewing is to formulate a few different clones, making different guesses as you compile them. For example, what if Summit really uses crystal 90°L instead of crystal 75°L? In that case, we'd need only about 1 pound (0.45 kg) of crystal (and slightly less pale malt than before) to hit the same color and gravity. Sometimes explicitly drawing out the differences will help you make your choices.

Finally, maybe the brewer would give you some feedback once you've compiled your clone.

In a worst-case scenario, one in which you've had to guess at many factors, you should still end up with a beer that is of the correct style even if it does taste distinctly different from your clone target. If your information was very complete, the success of your clone will rest mostly on your brewing skill and how closely the recipe's assumptions match the parameters of your system. (For an outline of the assumptions *BYO* uses, see page 157.)

JUST BREW IT

Of course, you can only really judge the success of your clone by brewing it. If you do brew the clone, taste it alone first and judge it against your memories of the target beer. Next, conduct a side-by-side tasting with the target beer. If the clone is significantly flawed, you will notice that the beer doesn't taste right even before the head-to-head comparison. If the clone is fairly good, you'll likely be pleased with the initial tasting. However, in the side-by-side tasting, you will pick up some differences. In the best-case scenario, your cloned beer will taste very similar to the target beer, even when directly compared to it.

Using your tasting comparison, you should be able to tweak the recipe to match the details of your system and move it from a "generic" clone to a "system-specific" clone for your brewery.

THE CLONE RECIPES

BYO RECIPE STANDARDIZATION

EXTRACT VALUES FOR MALT EXTRACT:
Liquid malt extract (LME) = 1.033–1.037
Dried malt extract (DME) = 1.045

EXTRACT EFFICIENCY: 65%
(i.e., 1 pound of 2-row pale malt—which has a potential extract value of 1.037 in one US gallon of water—would yield a wort of 1.024.)

POTENTIAL EXTRACT FOR GRAINS:

- 2-row base malts = 1.037–1.038
- wheat malt = 1.037
- 6-row base malts = 1.035
- Munich malt = 1.035
- Vienna malt = 1.035
- crystal malts = 1.033–1.035
- chocolate malts = 1.034
- dark roasted grains = 1.024–1.026
- flaked maize and rice = 1.037–1.038

HOPS:
We calculate IBUs based on 25 percent hop utilization for a 1-hour boil of hop pellets at specific gravities less than 1.050. For post-boil hop stands, we calculate IBUs based on 10 percent hop utilization for 30-minute hop stands at specific gravities less than 1.050. Increase the hop dosage by 10 percent if using whole-leaf hops.

ASSUMPTIONS:

- We use US gallons whenever gallons are mentioned.
- Unless otherwise specified, all mashes are 60 minutes and single infusion, with 1.25 quarts of water per pound of grain (2.6 L/Kg).
- Collect enough wort to boil 6.5 gallons (25 L) unless otherwise specified.
- Unless otherwise specified, the fermentation temperature matches the temperature to which you should chill the beer before pitching yeast.

ALASKAN BREWING COMPANY
SMOKED PORTER

(5 gallons/19 L, all-grain) OG = 1.065 FG = 1.015 IBU = 45 SRM = 58 ABV = 6.5%

Dark, robust, and smoky when young, this porter develops notes of sherry, Madeira, and raisin as it ages. It has a chewy malt character and is chocolaty with a smoky, oily finish.

INGREDIENTS

8.25 lb. (3.74 kg) 2-row pale malt
4 lb. (1.81 kg) Munich malt
12 oz. (0.34 kg) crystal malt (45°L)
11 oz. (0.31 kg) chocolate malt
7 oz. (0.20 kg) black patent malt
10.75 AAU Chinook hops (60 min.) (0.90 oz./25 g
   at 12% alpha acids)

3.75 AAU Willamette hops (15 min.) (0.75 oz./21 g
   at 5% alpha acids)
Wyeast 1968 (London ESB Ale) or White Labs WLP002
   (English Ale or LalBrew London ESB) yeast
3/4 cup (150 g) dextrose (for priming)

STEP BY STEP

Before brew day, smoke 1 pound (0.45 kg) of the Munich malt with alder wood. (Alaskan Brewing Company actually has its grains smoked at Taku Smokeries, a local Juneau smokehouse.) On brew day, mash the grains at 154°F (68°C). Mash out, vorlauf, and then sparge at 170°F (77°C). Boil for 90 minutes, adding the hops at the times indicated in the ingredients list. Pitch the yeast and ferment at 68°F (20°C) until final gravity is reached.

EXTRACT WITH GRAINS OPTION: Replace the 2-row pale and Munich malts with 2.25 pounds (1.0 kg) dried malt extract, 4.75 pounds (1.25 kg) liquid malt extract, and 1.25 pounds (0.56 kg) Munich malt. Smoke 1 pound (0.45 kg) of the Munich malt. Heat 4.7 quarts (4.4 L) of water to 165°F (74°C). Steep the crushed grains at 154°F (68°C) for 45 minutes. Rinse with 3.5 quarts (3.3 L) of water at 170°F (77°C). Add the dried malt extract and water to make 2.5 gallons (9.5 L) of wort. Boil for 60 minutes, adding the hops at the times indicated in the ingredients list and liquid malt extract with 15 minutes remaining. Cool the wort, transfer to your fermenter, and top up with filtered water to 5 gallons (19 L). Follow the remaining portion of the all-grain recipe.

TIPS FROM THE PROS

Dayton Canaday, Alaskan's plant manager, says, "When smoking, always use chlorine-free water to moisten the grain. Use a cool fire with no glowing coals or flames. Use a fine mesh to protect the malt from ash and tar. Ensure the malt is well-dried for best storage. You can check the flavor profile of your home-smoked malt by making a 'tea' from it. Pour 2 cups of hot water (140°F) over 1 cup of smoked malt. Allow this to steep for 10 minutes, then cool the tea to room temperature. Strain the tea and taste it to assess the smoke flavor. The flavor will be mild in the tea but more evident in the beer as the aromatics will be enhanced in the finished product due to the carbonation."

CHAPTER 3

THE ALCHEMIST
HEADY TOPPER

(5.5 gallons/21 L, all-grain) OG = 1.076 FG = 1.014 IBU = 100+ SRM = 6 ABV = 8%

The Alchemist specializes in fresh, unfiltered IPAs, and Heady Topper is the brewery's crown jewel. Featuring a proprietary blend of six hops, this beer boasts a complex and unique bouquet of hop flavor without any astringent bitterness.

INGREDIENTS

15 lb. (6.8 kg) British 2-row pale malt

6 oz. (170 g) Caravienne malt

1 lb. (0.45 kg) turbinado sugar (10 min.)

7 AAU Magnum hops (60 min.) (0.5 oz./14 g at 14% alpha acids)

13 AAU Simcoe hops (30 min.) (1 oz./28 g at 13% alpha acids)

5.75 AAU Cascade hops (0 min.) (1 oz./28 g at 5.75% alpha acids)

8.6 AAU Apollo hops (0 min.) (0.5 oz./14 g at 17.2% alpha acids)

13 AAU Simcoe hops (0 min.) (1 oz./28 g at 13% alpha acids)

10.5 AAU Centennial hops (0 min.) (1 oz./28 g at 10.5% alpha acids)

7 AAU Columbus hops (0 min.) (0.5 oz./14 g at 14% alpha acids)

1 oz. (28 g) Chinook hops (primary dry hop)

1 oz. (28 g) Apollo hops (primary dry hop)

1 oz. (28 g) Simcoe hops (primary dry hop)

1.25 oz. (35 g) Centennial hops (secondary dry hop)

1.25 oz. (35 g) Simcoe hops (secondary dry hop)

1 tbsp. polyclar

The Yeast Bay (Vermont Ale), White Labs WLP095 (Burlington Ale), Omega Yeast Labs (DIPA Ale) or Imperial Yeast A04 (Barbarian), or LalBrew New England yeast

$^2/_3$ cup (130 g) dextrose (if priming)

STEP BY STEP

Mash the grains at 153°F (67°C). Mash out, vorlauf, and then sparge at 170°F (77°C). Collect at least 6.5 gallons (25 L) of wort. Boil for 60 minutes, adding the hops as instructed. After the boil is complete, begin a whirlpool in the kettle and let the knockout hops rest in the hot wort for at least 30 minutes before chilling. Chill the wort, pitch the yeast, and ferment at 68°F (20°C) for 1 week. After final gravity has been achieved, add a clarifying agent such as polyclar. Three days later, add your first set of dry hops to the primary fermenter. After 7 days, rack the beer off the dry hops and yeast cake into a keg or secondary fermenter. Purge with carbon dioxide if available. Add the second set of dry hops to the keg or secondary fermenter. After 5 days, add the priming sugar and bottle or keg.

   EXTRACT WITH GRAINS OPTION: Substitute the British 2-row pale malt with 9.9 pounds (4.5 kg) light liquid malt extract and 1 pound (0.45 kg) extra light dried malt extract. Steep the crushed grains in 2 quarts (1.9 L) of water for 20 minutes at 155°F (68°C). Rinse the grain with hot water and add water to achieve 6.5 gallons (25 L) in your kettle. Turn off the heat, add the malt extract to your kettle, and stir until fully dissolved. Turn the heat back on and boil for 60 minutes, adding the hops as instructed. After the boil is complete, begin a whirlpool in the kettle and let the knockout hops rest in the hot wort for at least 30 minutes before chilling. Follow the remaining portion of the all-grain recipe.

TIPS FROM THE PROS

The goal is to get at least 5.5 gallons (21 L) into your fermenter to compensate for the loss of wort that will occur during dry hopping. Make sure your primary fermenter has enough headspace to accommodate that much wort plus a large kräusen. John Kimmich's biggest piece of advice for trying to clone this beer: "Technique and water treatment."

For extract brewers who use concentrated boils (usually around 3 gallons/11 L), try adding most of the extract near the end of the boil. This will keep the gravity low, which may help alpha acids convert into iso-alpha acids.

ALESMITH BREWING COMPANY
OLD NUMBSKULL

(5 gallons/19 L, all-grain) OG = 1.102 FG = 1.017 IBU = 90 SRM = 15 ABV = 11%

There are American barleywines, and then there are West Coast–style American barleywines–and if you like big beers, Old Numbskull doesn't disappoint. Packed with caramel and toffee notes, this copper-colored ale boasts an aggressive bitterness.

INGREDIENTS

16 lb. 13 oz. (7.6 kg) Gambrinus 2-row pale malt
16.3 oz. (0.46 kg) golden brown sugar
9.8 oz. (0.28 kg) Munich dark malt
6.5 oz. (0.19 kg) Crisp crystal malt (45°L)
4.9 oz. (0.14 kg) Crisp CaraMalt malt
4.9 oz. (0.14 kg) Gambrinus Honey malt
4.9 oz. (0.14 kg) Simpsons CaraMalt malt
4.9 oz. (0.14 kg) Simpsons CaraMalt Light malt
4.9 oz. (0.14 kg) flaked barley
16 AAU Chinook hops (FWH) (1.5 oz./42 g
  at 11% alpha acids)
15 AAU Columbus hops (FWH) (1.1 oz./31 g
  at 14% alpha acids)
9.8 AAU Warrior hops (FWH) (0.65 oz./18 g
  at 15% alpha acids)
2.3 AAU Simcoe hops (30 min.) (0.19 oz./5.4 g
  at 12% alpha acids)

1.1 AAU Chinook hops (15 min.) (0.1 oz./2.8 g
  at 11% alpha acids)
0.6 AAU Cascade hops (5 min.) (0.1 oz./2.8 g
  at 6% alpha acids)
1.7 AAU Palisade hops (2 min.) (0.28 oz./7.8 g
  at 6% alpha acids)
1 oz. (28 g) Amarillo hops (dry hop)
1 oz. (28 g) Columbus hops (dry hop)
1 oz. (28 g) Chinook hops (dry hop)
1 oz. (28 g) Simcoe hops (dry hop)
1 oz. (28 g) Palisade hops (dry hop)
1 oz. (28 g) Warrior hops (dry hop)
White Labs WLP001 (California Ale), Wyeast 1056
  (American Ale), or SafAle US-05 yeast
$^2/_3$ cup (130 g) dextrose (if priming)

STEP BY STEP

Mash the grains at 150°F (66°C). Boil for 90 minutes, adding the hops as indicated in the ingredients list. Pitch the yeast and ferment at 65 to 68°F (18 to 20°C) until final gravity is reached. Bottle or keg as usual.

EXTRACT WITH GRAINS OPTION: Omit the flaked barley. Reduce pale malt to 1 pound (0.45 kg). Add 8.66 pounds (3.9 kg) of light dried malt extract. Steep the grains at 150°F (66°C). Boil at least 4 gallons (15 L) of wort. Add the hops at the times indicated in the all-grain ingredients list. Reserve roughly half of the malt extract until 5 minutes before the end of the boil. Chill the wort, transfer to your fermenter, and top up with filtered water to 5 gallons (19 L). Follow the remaining portion of the all-grain recipe.

ALLAGASH BREWING COMPANY
ALLAGASH WHITE

(5 gallons/19 L, all-grain) OG = 1.049 FG = 1.010 IBU = 20 SRM = 3 ABV = 5.2%

One of America's most awarded Belgian-style witbiers, Allagash White features coriander and Curaçao orange peel that delivers a refreshing balance of citrus and spice.

INGREDIENTS

5.5 lb. (2.5 kg) 2-row pale malt
2 lb. (0.91 kg) malted red wheat
2 lb. (0.91 kg) raw white wheat
0.3 lb. (0.14 kg) dextrin malt
0.5 lb. (0.23 kg) flaked oats
3.9 AAU Nugget hops (60 min.) (0.3 oz./8.5 g
  at 13% alpha acids)
4.1 AAU Crystal hops (10 min.) (1.25 oz./35 g
  at 3.3% alpha acids)

0.75 oz. (21 g) Czech Saaz hops (0 min.)
0.25 oz. (7 g) Curaçao orange peel (0 min.)
0.5 oz. (14 g) coriander, crushed (0 min.)
Wyeast 3463 (Forbidden Fruit), White Labs WLP400
  (Belgian Wit Ale), Imperial Yeast B44 (Whiteout),
  or SafAle K-97 yeast
$^7/_8$ cup (175 g) dextrose (if priming)

STEP BY STEP

Heat 15.5 quarts (14.6 L) of strike water to 165°F (74°C). Mix with grains. The mash should stabilize at about 152°F (67°C). Hold at this temperature for 60 minutes, then raise temperature to mash out at about 168°F (76°C), by either infusion of boiling water, decoction, or other means. Vorlauf until wort runs clear, then begin the sparge process. Collect 7 gallons (26.5 L) and bring to a boil. Total boil time is 75 minutes, adding hops as indicated. After boil is complete, turn off the heat, add the final hop addition along with the coriander and orange peel (it helps to bag the coriander and orange peel), and give a long stir to create a whirlpool. After 15 minutes remove the spice bag.

Chill the wort to 65°F (18°C). There should be about 5.5 gallons (21 L) of wort in your fermenter. Add yeast and aerate wort if using liquid yeast. Place your fermenter in a temperature-stable place in the 68–72°F (20–22°C) range. Bottle or keg after fermentation is complete, targeting a carbonation level of 2.7 v/v.

PARTIAL MASH OPTION: Replace the 2-row pale with 3.3 pounds (1.5 kg) extra light dried malt extract and 4.75 pounds (1.25 kg) liquid malt extract. Put all the grains in a steeping bag, then heat 2 gallons (7.6 L) of water to 159°F (71°C). Add the grain bag. The target mash temperature is 152°F (67°C). Hold at this temperature for 60 minutes. Rinse the grain bag with 170°F (77°C) water to top up to 7 gallons (26.5 L), remove bag, then boil. (If your brew kettle doesn't allow for that large of a volume, rinse the grain bag with another gallon [4 L] of water, remove bag, then raise to a boil.) When boil is achieved, take the kettle off the flame and slowly add the extract while stirring. Return to heat source and boil for 60 minutes, adding hops as indicated. After boil is complete, turn off the heat, add the final hop addition along with the coriander and orange peel (it helps to bag the coriander and orange peel), and give a long stir to create a whirlpool. After 15 minutes remove the spice bag.

Chill the wort to 65°F (18°C). If you brewed a smaller volume, top off fermenter with prechilled water. In either case, the goal is to collect 5.5 gallons (21 L) of wort in your fermenter. Add yeast and aerate wort if using liquid yeast. Place your fermenter in a temperature-stable place in the 68–72°F (20–22°C) range. Bottle or keg after fermentation is complete, targeting a carbonation level of 2.7 v/v.

ANCHOR BREWING COMPANY
ANCHOR STEAM BEER

(5 gallons/19 L, all-grain) OG = 1.050 FG = 1.013 IBU = 30 SRM = 9 ABV = 4.9%

This beer takes its name from the days when beer was made in the cool climate of San Francisco on rooftops in the nineteenth century–open vessels were used to help cool the beer quickly. Steam is the beer that convinced Fritz Maytag to buy the brewery in 1965 and carry on the brewing tradition that started there in the late 1800s. It features a deep amber color and Northern Brewer hops.

INGREDIENTS

9 lb. 2 oz. (4.1 kg) 2-row pale malt
1 lb. 5 oz. (0.6 kg) caramel malt (40°L)
0.25 oz. (7 g) gypsum (optional if using
    very low mineral water)
4.8 AAU US Northern Brewer pellet hops (60 min.)
    (0.5 oz./14 g at 9.6% alpha acids)

2.4 AAU US Northern Brewer pellet hops (20 min.)
    (0.25 oz./7 g at 9.6% alpha acids)
0.5 oz. (14 g) US Northern Brewer pellet hops (0 min.)
    White Labs WLP810 (San Francisco Lager) or
    Wyeast 2112 (California Lager) or Mangrove
    Jack's M54 (California Lager) yeast
$^2/_3$ cup (130 g) dextrose (if priming)

STEP BY STEP

Mash the grains (with optional gypsum) at 149°F (65°C). Mash out, vorlauf, and then sparge with 3.33 gallons (12.6 L) of 168°F (76°C) water. Top up if necessary to obtain 6 gallons (23 L) of 1.041 SG wort. Boil the wort for 60 minutes, adding the hops according to the ingredients list. After the boil, turn off the heat and chill the wort to slightly below fermentation temperature, about 59°F (15°C). Pitch the yeast and ferment at 61°F (16°C) for 7 days before raising the temperature to 66°F (19°C) for 3 days for a diacetyl rest. Once the beer reaches final gravity (approximately 14 days total), bottle or keg the beer and carbonate. Store cold for approximately 2 weeks before serving.

EXTRACT WITH GRAINS OPTION: Substitute 6.25 pounds (2.8 kg) golden liquid malt extract for the 2-row pale malt. Place the milled grains in a muslin brewing bag and steep in 3 quarts (2.8 L) of 149°F (65°C) water for 15 minutes. Remove the grain and rinse with 1 gallon (3.8 L) of hot water. Add water and gypsum (if using) to reach a volume of 5.6 gallons (21.2 L) and heat to boiling. Turn off the heat, add the liquid malt extract, and stir until completely dissolved. Top up with filtered water if necessary to obtain 6 gallons (23 L) of 1.041 SG wort. Follow the remaining portion of the all-grain recipe.

ANDERSON VALLEY BREWING COMPANY
BLOOD ORANGE GOSE

(5 gallons/19 L, all-grain) OG = 1.038 FG = 1.005 IBU = 12 SRM = 3 ABV = 4.4%

Anderson Valley has become well-known for their variety of fruited Goses. This example uses blood orange juice that imparts tangy citrus notes that complement the Champagne-like flavors.

INGREDIENTS

5.5 lb. (2.5 kg) 2-row pale malt
2.4 lb. (1.1 kg) malted white wheat
~2 oz. (57 g) rice hulls
0.43 lb. (195 g) blood orange juice concentrate
3.3 AAU Nugget hops (60 min.) (0.25 oz./7 g
  at 13.1% alpha acids)
0.016 oz. (0.45 g) Indian coriander (fine ground) (5 min.)

0.61 oz. (17.2 g) sea salt
Lactobacillus culture, such as Wyeast 5335,
  White Labs WLP672, or WildBrew Sour Pitch
White Labs WLP029 (German Ale/Kölsch),
  Wyeast 1007 (German Ale), or SafAle K-97 yeast
3/4 cup (150 g) dextrose (if priming)

STEP BY STEP

Mash in at 150°F (66°C) with the grains and rice hulls. Rest 60 minutes and lauter as normal. As the kettle fills, begin to introduce an inert gas (usually nitrogen, but CO_2 works well too) into the top of the kettle. Stop runoff at 1.008. Once the wort is in the kettle, mix in cooled water to achieve a temperature of 118°F (48°C)–or recommended pitch temperature from the manufacturer–and a gravity of about 1.034. Add *Lactobacillus* propagation. Pitching rate is ~500 mL at 1 x 108 cells per mL (or approximately 5 x 1010 total). Allow to sour to desired pH (between 3.3-3.4). Hold at the recommended souring temperature. Once the pH is reached, boil the wort for 45 minutes, adding hops at beginning of the boil and the coriander at the end.

Pitch German ale yeast at 68-70°F (20-21°C). Add the blood orange juice concentrate near the end of active fermentation. At the end of fermentation, add the fully hydrated salt solution. Bottle and prime or keg and force carbonate as usual.

EXTRACT WITH GRAINS OPTION: Rice hulls are not needed. Swap out the pale and wheat malts for 2.2 pounds (1 kg) wheat dried malt extract and 2 pounds (0.91 kg) extra light dried malt extract. Heat 23 qts. (22 L) to 180°F (82°C) and stir in the dried malt extract. Hold at this temperature for 15 minutes for pasteurization, then cool wort to *Lacto* pitching temperature. Follow all-grain instructions for the remainder of the steps.

BELL'S BREWERY
HOPSLAM ALE

(5 gallons/19 L, all-grain) OG = 1.086 FG = 1.010 IBU = 65+ SRM = 7 ABV = 10%

Starting with six different hop varieties added to the brew kettle and culminating with a massive dry hop addition of Simcoe hops, Bell's Hopslam Ale possesses the most complex hopping schedule in the Bell's repertoire.

INGREDIENTS

11 lb. (5 kg) 2-row malt

5.5 lb. (2.5 kg) pale ale malt

0.5 lb. (0.23 kg) caramel malt (40°L)

12 oz. (0.34 kg) honey (0 min.)

10 oz. (285 g) dextrose (0 min.)

2.3 AAU Crystal or Mt. Hood hops (45 min.)
(0.5 oz./14 g at 4.5% alpha acids)

6 AAU Mosaic hops (20 min.) (0.5 oz./14 g
at 12% alpha acids)

2.3 AAU AAU Glacier or Fuggle hops (20 min.)
(0.5 oz./14 g at 4.5% alpha acids)

10 AAU Centennial hops (15 min.) (1 oz./28 g
at 10% alpha acids)

6 AAU Mosaic hops (5 min.) (0.5 oz./14 g
at 12% alpha acids)

2.3 AAU AAU Glacier or Fuggle hops (5 min.)
(0.5 oz./14 g at 4.5% alpha acids)

2 oz. (56 g) Amarillo hops (0 min.)

0.5 oz. (14 g) Crystal or Mt. Hood hops (0 min.)

4 oz. (113 g) Simcoe hops (dry hop)

White Labs WLP001 (California Ale), Wyeast 1056
(American Ale), or SafAle US-05 yeast

³/₄ cup (150 g) dextrose (if priming)

STEP BY STEP

Heat 22 qts. (21 L) of strike water to 163°F (73°C). Mix with grains; the mash should stabilize at about 149°F (65°C). Hold at this temperature for 70 minutes, then raise temperature to mash out at about 168°F (76°C) by either infusion of boiling water, decoction, or other means. Vorlauf until wort runs clear, then begin the sparge process. Collect approximately 7 gallons (26.6 L) and bring to a boil. The gravity of the wort at this point should be 1.068. Supplement with more dextrose or less if your gravity is off.

Total boil time is 75 minutes, adding hops as indicated. After boil is complete, turn off the heat, add the dextrose, honey, and final hop addition, then give a long stir to create a whirlpool. Let wort settle for 15 minutes.

Chill wort to 68–74°F (20–23°C). There should be about 5.5 gallons (21 L) of wort in your fermenter. Top fermenter up with cold water if you are short. Aerate wort and add yeast. Place your fermenter in a temperature-stable place that is in the 68–74°F (20–23°C) range. On day five of fermentation, add the dry hops addition. Bottle or keg after 3 days on dry hops complete.

EXTRACT WITH GRAINS OPTION: Replace the pale and pale ale malts with 6.6 pounds (3 kg) light liquid malt extract and 4 pounds (1.8 kg) extra light dried malt extract. Place crushed grains in a muslin bag and steep in 1 gallon (4 L) of water at 160°F (71°C) for 20 minutes in your brewpot. Remove the grains and fill up your kettle to 3.5 gallons (13 L) of water. Bring to a boil and then remove kettle from heat and add malt extracts. Stir until dissolved and return kettle to heat and boil for 45 minutes. Add hops as indicated.

When the boil is complete, add the honey, dextrose, and final hop addition, then give the wort a long stir to create a whirlpool. Let wort settle for 10 minutes. Chill the wort to 68–74°F (20–23°C). Fill your sanitized fermenter with 2 gallons (8 L) of cold water and transfer chilled wort to fermenter. Top fermenter up to 5.5 gallons (21 L) with cold water. Aerate wort and add yeast. Place your fermenter in a temperature-stable place that is in the 68–74°F (20–23°C) range. On day five of fermentation, add the dry hop addition either directly into the fermenter or by transferring beer on top of dry hops in a secondary fermenter. Bottle or keg after fermentation is complete.

BELL'S BREWERY
OBERON

(5 gallons/19 L, all-grain) OG = 1.056   FG = 1.012   IBU = 30   SRM = 5   ABV = 5.8%

Oberon is a wheat ale fermented with Bell's Brewery's signature house ale yeast, mixing a spicy hop character with mildly fruity aromas, often causing many to assume there is an addition of fruit or spices. The addition of wheat malt lends a smooth mouthfeel, making it a classic summer beer. Oberon is one of Bell's best-selling beers, even though it's only available from March–August each year.

INGREDIENTS

6 lb. (2.7 kg) 2-row malt
6 lb. (2.7 kg) white wheat malt
8 oz. (0.23 kg) Munich malt
8 oz. (0.23 kg) Carapils malt
3.5 AAU Hersbrucker hops (60 min.)
 (1 oz./28 g at 3.4% alpha acids)

3.5 AAU Hersbrucker hops (30 min.)
 (1 oz./28 g at 3.4% alpha acids)
2 oz. (56 g) Saaz hops (0 min.)
White Labs WLP001 (California Ale), Wyeast 1056
 (American Ale), or SafAle US-05 yeast
3/4 cup (150 g) dextrose (if priming)

STEP BY STEP

Heat 17.3 quarts (16.4 L) of strike water to 165°F (74°C). Mix with grains; the mash should stabilize at about 152°F (67°C). Hold at this temperature for 60 minutes, then raise temperature to mash out at about 168°F (76°C) by either infusion of boiling water, decoction, or other means. Vorlauf until wort runs clear, then begin the sparge process. Collect approximately 6.5 gallons (24.6 L) and bring to a boil. Total boil time is 60 minutes, adding hops as indicated.

After boil is complete, turn off the heat, add the final hop addition, then give a long stir to create a whirlpool. Let wort settle for 20 minutes, then chill the wort to 68–74°F (20–23°C). There should be about 5.5 gallons (21 L) of wort in your fermenter. Top fermenter up with cold water if you are short. Aerate wort and add yeast. Place your fermenter in a temperature-stable place that is in the 68–74°F (20–23°C) range. Bottle or keg after fermentation is complete.

EXTRACT WITH GRAINS OPTION: Replace the 2-row and wheat ale malts with 6.6 pounds (3 kg) wheat liquid malt extract and 1 pound (0.45 kg) extra light dried malt extract. Place crushed grains in a muslin bag and steep them in 1 gallon (4 L) of water at 160°F (71°C) for 20 minutes in your brewpot. Remove the grains and fill up your kettle to 3.5 gallons (13 L) of water. Bring to a boil and then remove kettle from heat and add malt extracts. Stir until dissolved and return kettle to heat and boil for 60 minutes. Add hops as indicated.

When the boil is complete, chill the wort to 68–74°F (20–23°C). Fill your sanitized fermenter with 2 gallons (8 L) of cold water and transfer chilled wort to fermenter. Top fermenter up to 5.5 gallons (21 L) with cold water. Aerate wort and add yeast. Place your fermenter in a temperature-stable place that is in the 68–74°F (20–23°C) range. Bottle or keg after fermentation is complete.

BIERSTADT LAGERHAUS
SLOW POUR PILS

(5 gallons/19 L, all-grain) OG = 1.047 FG = 1.012 IBU = 33 SRM = 3 ABV = 4.5%

Slow Pour Pils is aptly named due to the 5-minute duration that a proper, multi-step pour of it demands. Dry biscuit, crackery malt, and hints of honey support the white pepper and floral hops.

INGREDIENTS

9 lb. (4.08 kg) German Pilsner malt

0.5 lb. (0.23 kg) acidulated malt

4 AAU Hallertau Mittelfrüh hops (first wort hop)
  (1 oz./28 g at 4% alpha acids)

6 AAU Hallertau Mittelfrüh hops (30 min.)
  (1.5 oz./43 g at 4% alpha acids)

4 AAU Hallertau Mittelfrüh hops (0 min.)
  (1 oz./28 g at 4% alpha acids)

Wyeast 2124 (Bohemian Lager), White Labs WLP830
  (German Lager), or SafLager W-34/70 yeast

3/4 cup (150 g) dextrose (if priming)

STEP BY STEP

For authenticity, this all-grain recipe employs a step mash coupled with a single decoction. Mill the grains, then mix with 3 gallons (11.2 L) of 142°F (61°C) strike water to reach a protein rest of 131°F (55°C). Hold for 10 minutes before raising the mash temperature to 144°F (62°C). Hold this temperature for 30 minutes. Again, raise the mash temperature to 160°F (71°C) and hold there for 40 minutes. Finally, remove one-third of your mash, thin, and boil for 10 minutes before adding the decoction back to the main mash.

Vorlauf until your runnings are clear, and lauter. Sparge the grains with 4.5 gallons (17 L) and top up as necessary to obtain 6.5 gallons (25 L) of wort. Boil for 90 minutes, adding hops at the times indicated.

After the boil is complete, perform a 5-minute whirlpool and then rapidly chill the wort to slightly below fermentation temperature, which is 47°F (8.5°C) for this beer. Pitch yeast.

Maintain fermentation temperature of 47°F (8.5°C) for 3.5 weeks or until the completion of primary fermentation, whichever is later. Then, reduce temperature to 38°F (3.5°C) gradually by dropping the temperature one degree every 2 days and rest there for 1 week. Perform a similar drop in temperature until you reach 32°F (0°C) and lager for 3–4 weeks. Bottle or keg the beer and carbonate to approximately 2.5 volumes.

EXTRACT-ONLY OPTION: Replace the Pilsner and acidulated malts with 5.5 pounds (2.5 kg) of Pilsen dried malt extract and 2 tsp. 88% lactic acid. Bring 6 gallons (23 L) of water to a boil. At some point prior to boiling, add the dried extract and lactic acid

TIPS FOR SUCCESS

Aim for a massive 2 L starter (OG = 1.035) to provide enough healthy cells for a strong fermentation; Bierstadt recommends 210 million cells per mL.

BOULEVARD BREWING COMPANY
TANK 7

(5 gallons/19 L, all-grain) OG = 1.071 FG = 1.007 IBU = 38 SRM = 4 ABV = 8.5%

Tank 7 has become one of Boulevard's most recognized brands and a terrific example of a modern saison. The recipe got its start as a riff on Saison, a brand in the Smokestack Series, when brewmaster Steven Pauwels was working on Saison Brett. The Smokestack Saison recipe was tweaked to boost the strength from 7.5% to 8.5% ABV and the beer was dry hopped with Amarillo hops, which were relatively new at the time. The brewers were tasting the base for Saison Brett from Fermenter #7 and really dug what they were tasting. The "clean version" of Saison Brett (essentially Tank 7 bottle-conditioned with Brett and other conditioning yeast) became Tank 7 and the rest is history.

INGREDIENTS

9.25 lb. (4.2 kg) North American 2-row Pilsner malt
2.5 lb. (1.13 kg) North American white wheat malt
1.5 lb. (680 g) invert sugar
1.8 AAU Magnum hops (first wort hop) (0.15 oz./4 g at 12% alpha acids)
6 AAU Simcoe hops (60 min.) (0.5 oz./14 g at 12% alpha acids)

19.95 AAU Amarillo hops (5 min.) (2 oz./56 g at 10% alpha acids)
0.5 oz (14 g) Amarillo hops (dry hop)
Wyeast 3787 (Trappist Style High Gravity), White Labs WLP530 (Abbey Ale), Omega Yeast Labs OYL-028 (Belgian Ale W), or SafAle BE-256 yeast
1 cup (200 g) dextrose (if priming)

STEP BY STEP

This recipe uses reverse osmosis (RO) water. Adjust all brewing water to a pH of 5.5 using phosphoric acid. Add 1 tsp. calcium chloride to the mash. Mash the malts at 145°F (63°C) for 50 minutes, heat to 154°F (68°C) and hold for 25 minutes, then heat to 163°F (73°C) and hold for 15 minutes. Start recirculating wort. Raise the temperature to 168°F (76°C) for 15 minutes. Sparge slowly and collect 6.5 gallons (24.5 L) of wort. Add first wort hops and invert sugar when sparging is complete. Heat to boiling, and boil the wort for 90 minutes, adding hops at the times indicated in the recipe. Adjust OG post-boil with RO water as required.

Chill the wort to 62–64°F (17–18°C), pitch the yeast, and ferment until complete. Dry hops should be added when gravity is about 1.016. Cool to 32–34°F (0–1°C) and cold condition for 4 days. Rack the beer, prime and bottle condition, or keg and force carbonate.

EXTRACT WITH GRAINS OPTION: Replace the Pilsner and wheat malts with 4 pounds (1.8 kg) Pilsen dried malt extract and 2.3 pounds (1 kg) wheat dried malt extract. Heat 6.5 gallons (24.5 L) of water in your brew kettle to 180°F (82°C). Turn off the heat and add the malt extract and sugar and stir thoroughly to dissolve completely. You do not want to feel extract at the bottom of the kettle when stirring with your spoon. Turn the heat back on and bring to a boil. Add the first wort hops while raising to a boil.

Boil the wort for 90 minutes, adding hops at the times indicated in the recipe. Adjust OG post-boil with RO water as required. Chill the wort to 62–64°F (17–18°C), pitch the yeast, and ferment until complete. Dry hops should be added when gravity is about 1.016. Cool to 32–34°F (0–1°C) and cold condition for 4 days. Rack the beer, prime and bottle condition, or keg and force carbonate.

ORVAL TRAPPIST ALE

(5 gallons/19 L, all-grain) OG = 1.059 FG = 1.002 IBU = 33 SRM = 12 ABV = 6.2%

Orval pours orange-brown with a big, rocky head. The very spritzy levels of carbonation with a slight sour note and distinctive Brett character make the beer feel prickly on the tongue. Orval is dry and has little hop bitterness or flavor, although it is the only Trappist ale to be dry hopped. This recipe creates a beer similar to the Orval beer distributed to the US with a higher ABV.

INGREDIENTS

8.5 lb. (3.8 kg) Pilsner malt
1.3 lb. (0.6 kg) English light crystal malt (40°L)
2.2 lb. (1 kg) candi syrup (1°L)
4.4 AAU Hallertau-Hersbrücker hops (60 min.)
  (1.1 oz./31 g at 4% alpha acids)
3.3 AAU Styrian Goldings hops (60 min.)
  (0.66 oz./19 g at 5% alpha acids)
2.5 AAU Styrian Goldings hops (15 min.)
  (0.5 oz./14 g at 5% alpha acids)

0.3 oz. (8 g) Strisselspalt whole cone hops (dry hop)
0.25 oz. (7 g) Hallertau-Hersbrücker whole cone
  hops (dry hop)
¼ tsp. yeast nutrients
White Labs WLP510 (Bastogne Belgian Ale), Wyeast 3522
  (Belgian Ardennes), or LalBrew Abbaye Belgian Ale yeast
Wyeast 3112 (Brettanomyces bruxellensis) or White Labs
  WLP650 (Brettanomyces bruxellensis) yeast
1 cup (200 g) dextrose (if priming)

STEP BY STEP

Heat 3.6 gallons (13.5 L) of water to 163°F (73°C), stir in crushed grains, and mash at 150°F (65°C). Mash for 60 minutes, then raise grain bed temperature to 162°F (72°C). Hold for 15 minutes. Recirculate until wort is clear, then begin running wort off to kettle. Sparge with 170°F (77°C) water to collect roughly 7 gallons (26.5 L). Boil wort for 90 minutes, adding hops as indicated. Add yeast nutrients with 15 minutes left in boil and candi syrup at the end of boil. Cool wort down to 59°F (15°C), aerate, and pitch the Belgian yeast strain. Fermentation temperature can be slowly raised up to 72°F (22°C) to finish. Rack to secondary when active fermentation is complete and add Brettanomyces and dry hops. Let condition for 3 weeks at 59°F (15°C) before bottling. Condition warm for 3 weeks in bottles before serving.

EXTRACT WITH GRAINS OPTION: Replace the Pilsner malt with 4 pounds (1.8 kg) Pilsen dried malt extract. In a large soup pot, heat 5 gallons (19 L) of water to 163°F (73°C). Steep crushed grains in a grain bag as the water heats. When the temperature hits 170°F (77°C), remove the grains. Off heat, stir in dried malt extract until it is fully dissolved. Turn heat back on and bring wort to a boil. Add the first charge of hops and begin the 60-minute boil. Follow the rest of the all-grain recipe, being sure to top off to 5 gallons (19 L) after chilling the wort.

BREWERY OMMEGANG
HENNEPIN FARMHOUSE SAISON

(5 gallons/19 L, all-grain) OG = 1.070 FG = 1.008 IBU = 24 SRM = 5 ABV = 8%

Hennepin is Brewery Ommegang's flagship Belgian-style golden ale. Brewed with coriander, ginger root, sweet orange peel, and grains of paradise, this beer is full-bodied, hoppy, and crisp.

INGREDIENTS

8.75 lb. (4 kg) Belgian Pilsner malt
2.5 lb. (1.1 kg) Belgian 2-row pale malt
2 lb. (0.91 kg) light candi sugar
6.5 AAU Styrian Golding hops (60 min.)
 (1.3 oz./36 g at 5% alpha acids)
1.75 AAU Saaz hops (2 min.) (0.5 oz./14 g
 at 3.5% alpha acids)

1 tsp. Irish moss (15 min.)
0.33 oz. (9 g) dried ginger root (5 min.)
1.5 oz. (43 g) fresh sweet orange peel (5 min.)
0.33 oz. (9 g) crushed coriander seed (2 min.)
0.1 oz (3 g) grains of paradise (2 min.)
White Labs WLP550 (Belgian Ale), Wyeast 1214
 (Belgian Abbey), or Mangrove Jack's M47 (Belgian
 Abbey) yeast
$^3/_4$ cup (150 g) dextrose (for priming)

STEP BY STEP

Brewery Ommegang uses a multiple-step mash starting at 122°F (50°C) and ending at 152°F (67°C). Once your wort is collected, add Styrian Golding (bittering) hops and boil for 60 minutes, adding the other ingredients as specified. Pitch the yeast and ferment at 68°F (20°C). (Option: Pitch the yeast to the wort at 64°F/18°C and let the temperature rise during fermentation, as high as 77°F/25°C.) Bottle or keg your beer, age for 2–3 weeks, and enjoy!

EXTRACT WITH GRAINS OPTION: Replace the Belgian Pilsner and 2-row pale malts with 0.25 pounds (0.11 kg) Muntons light dried malt extract and 6.6 pounds (2.97 kg) Muntons light liquid malt extract (late addition). Steep the crushed grains in 3 quarts (2.8 L) of water at 150°F (66°C) for 45 minutes. Rinse the grains with 1.5 quarts (about 1.5 L) of water at 170°F (77°C). Add water to make 3 gallons (11 L). Add the dried malt extract and sugar and bring the wort to a boil. Add the Styrian Golding (bittering) hops and boil for 60 minutes, adding the other ingredients as specified. When done boiling, cool the wort rapidly, transfer to a sanitized fermenter, top up to 5 gallons (19 L) with filtered water, and pitch the yeast. Follow the remaining portion of the all-grain recipe.

BROUWERIJ DUVEL MOORTGAT
DUVEL

(5 gallons/19 L, all-grain) OG = 1.061   FG = 1.007   IBU = 30   SRM = 3   ABV = 8.5%

Duvel is the classic Belgian golden ale. Although it's strong (8.5% ABV), the beer is extremely light in color and dry in taste.

INGREDIENTS

11.5 lb. (5.2 kg) Belgian Pilsner malt
0.5 lb. (0.23 kg) dextrose (15 min.)
6 AAU Styrian Golding hops (60 min.)
  (1.2 oz./34 g at 5% alpha acids)
4 AAU Saaz hops (15 min.) (1 oz./28 g at 4% alpha acids)
0.75 oz. (21 g) Saaz hops (0 min.)
$^1/_4$ tsp. yeast nutrients (15 min.)

1 lb. (0.45 kg) dextrose (dosage)
$^1/_{16}$ tsp. ("a pinch") yeast nutrients (dosage)
1 tsp. Irish moss (15 min.)
Wyeast 1388 (Belgian Strong Ale), White Labs WLP570 (Belgian Golden Ale), or Mangrove Jack's M31 (Belgian Tripel) yeast
1 cup (200 g) dextrose (for priming)

STEP BY STEP

In your brew kettle, mash in to 131°F (55°C) and heat the mash slowly, over 15 minutes, to 140°F (60°C). Add boiling water to raise the temperature to 148°F (64°C) and hold for 60 minutes. Mash out, vorlauf, and then sparge at 168°F (76°C). (Option: To increase wort fermentablility, mash in at 99°F [37°C] and slowly ramp up the temperature to mash-out temperature. The ramp time can take anywhere from 90 minutes to as long as 3 hours.) Collect enough wort in the brew kettle to compensate for a longer boil. Boil for 90 minutes, adding the hops as indicated. At the end of the boil, add the last charge of hops, rapidly cool the wort, and transfer it to a sanitized fermenter. The brewers of Duvel pitch at 60°F (16°C) and let the temperature rise to as high as 84°F (29°C) during primary fermentation. If you are not capable of ramping the fermentation temperature, pitch the yeast and then ferment at 68°F (20°C). When primary fermentation is slowing down (3–4 days), add the dosage sugar with a pinch of yeast nutrients that have been dissolved in hot water held at 160°F (71°C) for 15 minutes. After final gravity is achieved, rack to a secondary fermenter and allow the beer to condition for 2–3 weeks. Bottle when the beer falls clear.

TIPS FROM *BYO*

The original gravity for the beer is the estimated wort gravity on brew day, while the ABV is calculated based on the original wort strength plus the sugar addition to the fermenter. The yeast will not have an easy job fully attenuating the wort. Help them out first by taking your time with the mash—longer mashes promote better attenuation. Second, make a proper yeast starter so you have a strong cell count at pitching. Finally, be patient with the yeast. It may take a bit longer than normal for this strain to fully attenuate.

EXTRACT-ONLY OPTION: Substitute the Pilsner malt with 6.3 (3.9 kg) Pilsen dried malt extract. In a large brewpot, heat 4 gallons (15 L) up to a boil. Remove from heat and stir in the dried malt extract. Bring back to a boil, then add the first charge of hops and begin the 60-minute boil. With 15 minutes left in boil, add the kettle sugar and second charge of hops and Irish moss, then turn off heat and stir in the liquid malt extract. Stir well to dissolve the extract, then resume heating. (Keep the boil clock running while you stir.) At the end of the boil, add the last charge of hops, cool the wort, and transfer to your fermenter. Top off the fermenter with filtered water to make 5 gallons (19 L). Follow the remaining portion of the all-grain recipe.

DESCHUTES BREWERY
BLACK BUTTE PORTER

(5 gallons/19 L, all-grain) OG = 1.053 FG = 1.013 IBU = 30 SRM = 29 ABV = 5.2%

This is the porter of record in the Pacific Northwest and the beer that Deschutes built their brewing empire upon. Every sip has layered notes of chocolate and coffee, enhanced by the creamy mouthfeel.

INGREDIENTS

10 lb. (4.5 kg) 2-row pale malt
9 oz. (252 g) chocolate malt
10 oz. (0.28 kg) crystal malt (60°L)
1 tsp. Irish moss (15 min.)
4 AAU Galena pellet hops (60 min.)
  (0.33 oz./9 g at 12% alpha acids)

4.25 AAU Cascade hops (30 min.) (0.85 oz./24 g
  at 5% alpha acids)
4.5 AAU Tettnanger hops (5 min.) (1.1 oz./31 g
  at 4% alpha acids)
Wyeast 1318 (London Ale III) or Lalbrew Verdant IPA yeast
³/₄ cup (150 g) dextrose (for priming)

STEP BY STEP

Mash the grains at 152°F (67°C) for 45 minutes. Mash out, vorlauf, and then sparge at 170°F (77°C) to get 6 gallons (23 L) of wort. Add 0.5 gallons (1.9 L) of water and boil for 60 minutes, adding the hops at times indicated in the ingredients list. Add the Irish moss with 15 minutes left in the boil. Chill the wort rapidly to about 65°F (18°C) and pitch the yeast. Ferment at 67°F (19°C) until final gravity is reached (about 7 to 10 days), then transfer to a secondary vessel or rack directly into bottles or keg. Let it age for at least 2 weeks.

EXTRACT WITH GRAINS OPTION: Substitute the 2-row malt with 2.5 pounds (1.13 kg) light dried malt extract and 3.3 pounds (1.5 kg) light liquid malt extract (late addition). Place the crushed malts in a nylon steeping bag and steep in 3 quarts (2.8 L) of water at 150°F (66°C) for 30 minutes. Rinse the grains with 1.5 quarts (about 1.5 L) of water at 170°F (77°C). Add water to make 3 gallons (11 L), then stir in the dried malt extract and bring to a boil. Boil for 30 minutes. Add the Galena pellet hops. Boil 30 minutes and add the Cascade hops. Boil 25 minutes and then add the Tettnanger hops. Boil for 5 minutes and remove from the heat. Add the liquid malt extract and Irish moss with 15 minutes left in the boil. Chill the wort, transfer to your fermenter, and top up with filtered water to 5 gallons (19 L). Follow the remaining portion of the all-grain recipe.

DEVILS BACKBONE BREWING COMPANY
VIENNA LAGER

(5 gallons/19 L, all-grain) OG = 1.050 FG = 1.011 IBU = 18 SRM = 10 ABV = 5.1%

The gold standard for Vienna lagers in North America, known for its smooth, malty finish and drinkability. This beer from Devils Backbone (Roseland, Virginia) has won gold at the Great American Beer Festival four times since 2012, plus numerous other accolades.

INGREDIENTS

4.1 lb. (1.9 kg) Pilsner malt
4.1 lb. (1.9 kg) Vienna malt
1.25 lb. (0.57 kg) dark Munich malt
1.25 lb. (0.57 kg) Weyermann Caraamber malt (26°L)
3.75 AAU German Northern Brewer hops (60 min.)
  (0.5 oz./14 g at 7.5% alpha acids)

1.75 AAU Czech Saaz hops (20 min.) (0.5 oz./14 g
  at 3.5% alpha acids)
Imperial L17 (Harvest), Omega Yeast OLY-114 (Bayern
  Lager), or Mangrove Jack's M76 (Bavarian Lager) yeast
¾ cup (150 g) dextrose (if priming)

STEP BY STEP

With a loose 3-to-1 water-to-grist ratio (~1.5 qts./lb.), conduct a protein rest at 125°F (52°C), holding for 30 minutes. Raise temperature to 147°F (64°C) and hold for 30 minutes for beta amylase rest. Raise temperature to 162°F (72°C) for 30 more minutes to convert alpha amylase. Mash out for 10 minutes at 170°F (77°C). Recirculate until clear, then sparge until you collect about 7 gallons (26.5 L) of wort. Boil for 90 minutes, adding hops as indicated. If you choose to add clarifiers such as Whirlfloc or Irish moss, do so with 15 minutes remaining in the boil.

Chill to 52°F (11°C), then pitch ample amount of healthy yeast. Oxygenate thoroughly. Allow temperatures to rise to 54–54.5°F (12°C) during the main fermentation period. When fermentation is about two-thirds complete, raise temperatures to 56°F (13°C) to finish fermenting remaining sugars. At end of fermentation, let rise to 57°F (14°C) for diacetyl rest. Seven to 10 days after pitching. Start cooling 2°F (1°C) per day until you reach a temperature between 42–44°F (6–7°C). Rack off the yeast and transfer to a secondary for aging. Crash cool to lagering temperatures between 28–34°F (-2°C to 1°C). Lager for a minimum of 2 weeks, but the beer will benefit from a lagering period of 4 weeks or more. Force carbonate to 2.5 volumes or prime and bottle condition.

PARTIAL MASH OPTION: Replace the Pilsner malt and 2.1 pounds (1 kg) of the Vienna malt for 2.3 pounds (1 kg) Pilsen dried malt extract and 1 pound (0.45 kg) Briess Goldpils Vienna dried malt extract. Heat 2 gallons (7.6 L) of water to 157°F (69°C) and place large steeping bag containing the crushed grains into a 5-gallon (19 L) pot. Submerge the bag and stir grains to ensure sufficient hydration. Mash for 45 minutes targeting a mash temperature of 149°F (65°C). Remove bag from pot and wash grains with enough 170°F (77°C) water to collect 2.5 gallons (9.5 L) of wort.

Top off kettle with water to make 3 gallons (11 L) and add 1 pound (0.45 kg) Pilsen dried malt extract to improve hop isomerization. Boil wort for 60 minutes, adding hops according to the schedule. Add remaining malt extract in the last 15 minutes of the boil. If you choose to add clarifiers such as Whirlfloc or Irish moss, do so with 15 minutes remaining in the boil. Chill to 52°F (11°C), then top off wort with prechilled water to bring volume up to 5 gallons (19 L). Pitch ample amount of healthy yeast. Oxygenate thoroughly. Allow temperatures to rise to 54–54.5°F (12°C) during the main fermentation period. When fermentation is about two-thirds complete, raise temperatures to 56°F (13°C) to finish fermenting remaining sugars. At end of fermentation, let rise to 57°F (14°C) for diacetyl rest. Follow the lagering and packaging instructions outlined in the all-grain version of this recipe.

DEVIL'S PURSE BREWING COMPANY
HANDLINE KÖLSCH

(5 gallons/19 L, all-grain) OG = 1.048 FG = 1.009 IBU = 20 SRM = 4 ABV = 5.1%

The flagship beer from the Cape Cod, Massachusetts, brewery, this crisp and refreshing beer is good anytime but best enjoyed while out on the beach or fishing. Brewed with American Vanguard hops, which is a descendent of German Hallertau, sets this rendition apart from its continental counterparts.

INGREDIENTS

7.25 lb. (3.29 kg) Pilsner malt
2.75 lb. (1.25 kg) Vienna malt
5 AAU Vanguard hops (60 min.) (1 oz./28 g at 5% alpha acids)

2.5 AAU Vanguard hops (5 min.) (0.5 oz./14 g at 5% alpha acids)
Wyeast 2565 (Kölsch), White Labs WLP003 (German Ale II), or SafAle K-97 yeast
³/₄ cup (150 g) dextrose (if priming)

STEP BY STEP

Start with either soft water, reverse osmosis (RO) water, or distilled water. Mill the grains, then mix with strike water at a 1.25 quarts/pound (2.6 L/kg) water-to-grain ratio, or 3.1 gallons (11.8 L) of 165°F (74°C) strike water to achieve a single-infusion rest temperature of 148°F (64°C). Adjust water with calcium salts and 88 percent lactic acid to stabilize mash pH at 5.3. Hold at this temperature for 60 minutes. Mash out to 170°F (77°C) if desired.

Vorlauf until your runnings are clear before directing them to your boil kettle. Batch or fly sparge the mash and runoff to obtain 6.5 gallons (25 L) of wort. Boil for 60 minutes, adding hops at the times indicated in the ingredients list. At 15 minutes left in boil, you may want to add either Irish moss or Whirlfloc as fining agents. After the boil, rapidly chill the wort to 65°F (18°C), then pitch a healthy amount of yeast, higher than your standard ale pitch rate. Maintain this temperature during active fermentation in order to prevent too much yeast character. Once primary fermentation is complete, and the beer has settled, you can bottle or keg the beer and carbonate to approximately 2.6 volumes. Alternatively, if you can, cold lager the beer for about 1 month.

EXTRACT-ONLY OPTION: Use 4 pounds (1.8 kg) Pilsen dried malt extract and 1.5 pounds (0.68 kg) Briess Goldpils Vienna malt extract. Bring 6.5 gallons (25 L) of water to roughly 150°F (66°C). Add the malt extracts, with stirring, before heating to a boil. The warmer water will help to dissolve the extracts. Boil for 60 minutes, adding hops at the indicated times left in the boil. At 15 minutes left in the boil, you may want to add either Irish moss or Whirlfloc as fining agents. After the boil, rapidly chill the wort to 65°F (18°C). Pitch yeast. Maintain fermentation temperature in order to prevent too colorful of a fermentation character. Once primary fermentation is complete, and the beer has settled, you can bottle or keg the beer and carbonate to approximately 2.6 volumes. Alternatively, if you can, cold lager the beer for about 1 month.

DREKKER BREWING COMPANY
HYPER SCREAM

(5 gallons/19 L, all-grain) OG = 1.076 FG = 1.013 IBU = 25 SRM = 6 ABV = 8.4%

This imperial New England–style IPA has a silky mouthfeel from generous oat and spelt additions, and a massive hop aroma due to an abundance of late-hop additions and kveik yeast.

INGREDIENTS

9 lb. (4.1 kg) Rahr 2-row barley malt
3.25 lb. (1.5 kg) Crisp naked oat malt
2.25 lb. (1 kg) Weyermann Carafoam malt
1.25 lb. (0.57 kg) Weyermann spelt (dinkel) malt
0.25 lb. (113 g) Simpsons Golden Naked Oats malt
3 AAU Magnum hops (60 min.) (0.25 oz./7 g
  at 12% alpha acids)
12 AAU Vic Secret hops (0 min.) (0.75 oz./21 g
  at 16% alpha acids)
9 AAU Citra hops (0 min.) (0.75 oz./21 g at 12% alpha acids)

1.5 oz. (42 g) Vic Secret hops (hopstand)
1.5 oz. (42 g) Citra hops (hopstand)
3.75 oz. (106 g) Vic Secret hops (1st dry hop)
3.75 oz. (106 g) Citra hops (1st dry hop)
1.75 oz. (50 g) Vic Secret hops (2nd dry hop)
1.75 oz. (50 g) Citra hops (2nd dry hop)
Omega Yeast OYL061 (Voss Kveik), Imperial Yeast A43
  (Loki), or LalBrew Voss Kveik
¾ cup (150 g) dextrose (if priming)

STEP BY STEP

This is a single-step infusion mash at 152°F (67°C) for 60 minutes. Vorlauf and sparge as usual to collect 6.5 gallons (24.6 L) of wort. Boil for 60 minutes, adding the first hop addition as the wort comes to a boil. Add the flameout hops at the end of the boil. Create a whirlpool and let settle for 10 minutes. Then cool the wort to 175°F (79°C) and add the hopstand addition and whirlpool for 10 minutes before cooling down to 80°F (27°C). Aerate, pitch yeast, and let temperature free rise to 100°F (38°C) during fermentation. The first dry hop additions should be added at about 24 hours at peak of fermentation. The second dry hop additions should be added on day five. Keep the beer on the hops for 3 more days, then package as normal. Kegging is preferred.

PARTIAL MASH OPTION: Reduce the Carafoam malt to 2 pounds (0.9 kg) and replace all the 2-row malt with 5 pounds (2.27 kg) extra light dried malt extract. Place all the crushed grains in a large grain bag. Heat 9.5 quarts (9 L) of water to 167°F (75°C), then submerge the grains into the water. Mix well and the mash should settle at 152°F (67°C). Try to maintain this temperature for 60 minutes. Remove the grain bag and rinse the grains with 2 gallons (7.6 L) of hot water. Top off the kettle to 6 gallons (23 L), then stir in the dried malt extract. Once all the extract is dissolved, bring wort up to a boil. Boil for 60 minutes, adding the first hop addition as the wort comes to a boil. Add the flameout hops at the end of the boil. Create a whirlpool and let settle for 10 minutes. Then cool the wort to 175°F (79°C) and add the hopstand addition and whirlpool for 10 minutes before continuing to cool down to 80°F (27°C). Follow the remainder of the all-grain recipe instructions.

FIRESTONE WALKER BREWING COMPANY
PIVO PILS

(5 gallons/19 L, all-grain) OG = 1.046 FG = 1.009 IBU = 40 SRM = 3 ABV = 5%

Inspired by the dry-hopped Tipopils from Birrificio Italiano, Firestone Walker's brewmaster Matt Brynildson drew upon multiple European influences to create this Pilsner that helped spawn a legion of new craft Pilsners across the US. It features floral aromatics, spicy herbal nuances, and lemongrass notes from German Saphir hops.

INGREDIENTS

9 lb. 6 oz. (4.3 kg) Weyermann Pilsner malt
8.3 AAU German Magnum hops (60 min.)
   (0.75 oz./21 g at 11% alpha acids)
2.3 AAU German Spalt Select hops (30 min.)
   (0.5 oz./14 g at 4.5% alpha acids)
0.8 oz. (22 g) German Saphir hops (0 min.)

0.8 oz. (22 g) German Saphir hops (dry hop)
7 g calcium chloride (if using reverse osmosis water)
White Labs WLP830 (German Lager), Wyeast 2124
   (Bohemian Lager), or SafLager W-34/70 yeast
³/₄ cup (150 g) dextrose (if priming)

STEP BY STEP

Mash the grains (and calcium chloride, if using) at 145°F (63°C) for 15 minutes. Raise the mash temperature to 155°F (68°C) and hold for 30 minutes. Mash out, vorlauf, and then sparge at 168°F (75°C). Collect 6 gallons (23 L) of 1.038 wort. Boil for 60 minutes, adding hops according to the ingredients list. Turn off the heat and chill the wort to slightly below fermentation temperature, about 48°F (9°C). Aerate the wort with pure oxygen or filtered air and pitch the yeast. Ferment at 50°F (10°C). After 4 days of fermentation, add the dry hop addition. After 7 days total, slowly raise the temperature to 60°F (16°C) for 3 days for a diacetyl rest, then slowly lower the beer to 34°F (1°C). Once at final gravity (approximately 14 days total), bottle or keg the beer and carbonate. Lager at 34°F (1°C) for approximately 1 month before serving.

EXTRACT WITH GRAINS OPTION: Replace the Pilsner malt with 6.5 pounds (2.9 kg) Pilsen liquid malt extract. Bring 5.5 gallons (21 L) of water and calcium chloride (if using) to boil, turn off the flame, and stir in the liquid malt extract until completely dissolved. Top up if necessary to obtain 6 gallons (23 L) of 1.038 SG wort. Boil for 60 minutes, adding the hops according to the ingredients list. Turn off the heat and chill the wort to slightly below fermentation temperature, about 48°F (9°C). Aerate the wort with pure oxygen or filtered air and pitch the yeast. Follow the remaining portion of the all-grain recipe.

TIPS FROM THE PROS

The biggest key to making this beer (and any good Pilsner) is to start with soft water. Firestone Walker runs all their brewing liquor through a reverse osmosis system before adding back calcium to reach 100 ppm for yeast health and to avoid beer scale formation on equipment. This is done with calcium chloride for malt-focused beers and with equal parts calcium chloride and calcium sulfate for hop-focused beers.

FIRESTONE WALKER BREWING COMPANY
VELVET MERLIN

(5 gallons/19 L, all-grain) OG = 1.061 FG = 1.020 IBU = 29 SRM = 44 ABV = 5.6%

This seasonal oatmeal stout is named in honor of Firestone Walker's brewmaster Matt "Merlin" Brynildson, who earned the nickname because of his magical ability to rack up top honors at prestigious beer competitions. This is a rich beer, with dark chocolate and roasted coffee flavors. It boasts a truly creamy mouthfeel and dry finish.

INGREDIENTS

8.75 lb. (4 kg) Rahr Standard 2-row malt
1 lb. 9 oz. (0.71 kg) flaked oats
1 lb. 9 oz. (0.71 kg) Briess roasted barley (300°L)
14 oz. (400 g) caramel malt (120°L)
5 oz. (140 g) Carapils malt
4 oz. (113 g) caramel malt (80°L)
4 oz. (113 g) Weyermann Carafa Special III malt
3.8 AAU Fuggle hops (60 min.) (0.85 oz./24 g
    at 4.5% alpha acids)
3.8 AAU Fuggle hops (30 min.) (0.85 oz./24 g
    at 4.5% alpha acids)
7 g calcium chloride (if using reverse osmosis water)
White Labs WLP002 (English Ale), Wyeast 1968
    (London ESB Ale), or LalBrew London ESB yeast
3/4 cup (150 g) dextrose (if priming)

STEP BY STEP

Mash the grains (and calcium chloride) at 145°F (63°C) for 15 minutes. Raise the mash temperature to 155°F (68°C) and hold for 30 minutes. Raise the temperature to a mash out of 168°F (76°C). Mash out, vorlauf, and then sparge at 168°F (76°C). You should now have 6 gallons (23 L) of 1.051 SG wort. Boil for 60 minutes, adding the hops at the times indicated. Chill the wort to slightly below fermentation temperature, about 66°F (19°C). Aerate the wort with pure oxygen or filtered air and pitch the yeast. Ferment at 68°F (20°C) until final gravity is reached. Bottle or keg as usual.

PARTIAL MASH OPTION: Substitute 5 pounds (2.27 kg) golden liquid malt extract and 1.5 pounds (0.68 kg) US 2-row pale malt in place of the Rahr Standard 2-row malt in the all-grain recipe. Place the milled 2-row pale malt and flaked oats in a muslin bag and steep in 10 quarts (9.5 L) of 149°F (65°C) water for 45 minutes. Remove the grains and rinse with 1 gallon (3.8 L) of hot water. Add the remaining crushed grains in a separate muslin bag and steep an additional 15 minutes. Add water to reach a volume of 5.4 gallons (20.4 L) and heat to boiling. Turn off the heat, add the liquid malt extract and optional calcium chloride, and stir until completely dissolved. Top up to obtain 6 gallons (23 L) of 1.051 SG wort. Boil for 60 minutes, adding hops according to the ingredients list. Follow the remaining portion of the all-grain recipe.

TIPS FROM THE PROS

If you're brewing the all-grain version of this recipe, keep in mind that oats contain many large beta-glucan gums and undegraded proteins as well. While this is great for adding mouthfeel, a large percentage of oats can contribute to a stuck sparge. At 12 percent flaked oats in this recipe, adding a handful of rice hulls is a good idea if you often experience sticky mashes. Add them prior to starting your lauter.

FORGOTTEN BOARDWALK BREWING COMPANY
FUNNEL CAKE

(5 gallons/19 L, all-grain) OG = 1.054 FG = 1.014 IBU = 26 SRM = 3 ABV = 5.4%

Forgotten Boardwalk Brewing Company, located in Cherry Hill, New Jersey, describes their Funnel Cake sweet cream ale as smelling like "old-fashioned Nilla wafers" and tasting like classic Jersey Shore boardwalk funnel cake.

INGREDIENTS

9 lb. (4.1 kg) Pilsner malt
1.25 lb. (0.57 kg) flaked oats
0.5 lb. (0.23 kg) light Munich malt (6°L)
0.25 lb. (0.11 kg) flaked maize
0.25 lb. (0.11 kg) lactose sugar (60 min.)
0.5 oz. (14 g) vanilla beans (dry hop)

7 AAU Magnum hops (60 min.) (0.5 oz./14 g
  at 14% alpha acids)
Wyeast 1056 (American Ale), White Labs WLP001
  (California Ale), or SafAle US-05 yeast
3/4 cup (150 g) dextrose (if priming)

STEP BY STEP

Mill the grains and mix with 3.5 gallons (13 L) of 164°F (73°C) strike water to reach a mash temperature of 152°F (67°C). Hold this temperature for 60 minutes. Vorlauf until your runnings are clear. Sparge the grains with 4.3 gallons (16.2 L) and top up as necessary to obtain 6 gallons (23 L) of wort. Boil for 60 minutes, adding lactose and hops at the start of the boil. Conduct a 15-minute whirlpool after flameout and then chill the wort to slightly below fermentation temperature, about 65°F (18°C). Aerate the wort with pure oxygen or filtered air and pitch yeast. Ferment at 66°F (19°C) for 7 days, then raise the temperature to 70°F (21°C) for 72 hours. After completion of fermentation, add the vanilla beans and rest for 3 days. Crash the beer to 35°F (2°C) for 48 hours, then bottle or keg the beer and carbonate to approximately 2.5 volumes.

PARTIAL MASH OPTION: Reduce the Pilsner malt to 0.5 pound (0.23 kg) and the flaked oats to 1 pound (0.45 kg). You will need 4.5 pounds (2 kg) Pilsen dried malt extract. Bring 1 gallon (3.8 L) of water to approximately 162°F (72°C) and hold there. Place the crushed grains in a large grain bag and submerge in the water, being sure to mix the grains well to avoid dough balls. Hold for 45 minutes, maintaining a temperature above 150°F (66°C) but below 162°F (72°C). Remove the grain bag and place in a colander. Wash the grain bag with 1 gallon (4 L) of hot water. Add dried malt extract while stirring, and stir until completely dissolved, then top off to 6 gallons (23 L). Follow the remaining portion of the all-grain recipe.

TIPS FOR SUCCESS

The light Munich, flaked oats, and flaked maize create a rich pastry background, while the lactose and vanilla (a light touch of each) provide just enough sweetness. If you find your version is a bit cloying, just bump up the bittering hop addition by a few AAUs.

FOUNDER'S BREWING COMPANY
BREAKFAST STOUT

(5 gallons/ 19 L, extract with grains) OG = 1.078 FG = 1.020 IBUs = 60 SRM = 59 ABV = 7.5%

Founder's describes this as "the coffee lover's consummate beer." Brewed with flaked oats, bitter and imported chocolate, and two types of coffee, this is indeed like the strong, dark cup of joe you'll want for breakfast–or anytime!

INGREDIENTS

13.2 lb. (6 kg) 2-row pale malt
22 oz. (0.62 kg) flaked oats
1 lb. (0.45 kg) chocolate malt (350°L)
12 oz. (0.34 kg) roast barley malt (450°L)
9 oz. (0.25 kg) debittered black malt (530°L)
7 oz. (0.19 kg) crystal malt (120°L)
14.3 AAU Nugget pellet hops (60 min.) (1.1 oz./31 g at 13% alpha acids)
2.5 AAU Willamette pellet hops (30 min.) (0.5 oz./14 g at 5% alpha acids)
2.5 AAU Willamette pellet hops (0 min.) (0.5 oz./14 g at 5% alpha acids)

2 oz. (57 g) ground Sumatran coffee
2 oz. (57 g) ground Kona coffee
2.5 oz. (71 g) dark, bittersweet chocolate
1.5 oz. (43 g) unsweetened chocolate baking nibs
1/2 tsp. yeast nutrient (15 min.)
1/2 tsp. Irish moss (15 min.)
White Labs WLP 001 (California Ale), Wyeast 1056 (American Ale), or SafAle US-05 yeast
3/4 cup (150 g) dextrose (for priming)

STEP BY STEP

Mash the crushed grains with 3.75 gallons (14 L) of water at 155°F (68°C) for 60 minutes. Vorlauf, then sparge slowly with 175°F (79°C) water. Add the hops and Irish moss according to times indicated in the ingredients list. Add the Sumatran coffee and two chocolate varieties at the end of the boil. Pitch the yeast and ferment at 68°F (20°C) until final gravity is reached. Transfer to a carboy, avoiding any splashing. Add the Kona coffee and condition for 1 week, then bottle or keg. Carbonate and age for 2 weeks.

EXTRACT WITH GRAINS OPTION: Substitute the 2-row pale malt with 6.6 pounds (3.0 kg) Briess light, unhopped malt extract, and 1.7 pounds (0.77 kg) light dried malt extract. Steep the crushed grain in 2 gallons (7.6 L) of water at 155°F (68°C) for 30 minutes. Remove grains from the wort and rinse with 2 quarts (1.8 L) of hot water. Add the liquid and dried malt extracts and bring to a boil. Add the hops and Irish moss as per the ingredients list. Add the Sumatran coffee and two chocolate varieties at the end of the boil. Add the wort to 2 gallons (7.6 L) of filtered water in a sanitized fermenter and top up to 5 gallons (19 L). Follow the remaining portion of the all-grain recipe.

FULLER, SMITH & TURNER PLC
FULLER'S ESB

(5 gallons/19 L, all-grain) OG = 1.060 FG = 1.014 IBU = 35 SRM = 15 ABV = 5.9%

This is a clone of the bottled version of Fuller's ESB (5.9% ABV), the beer available in the United States. There is also a cask version (at 5.5% ABV) available in England.

INGREDIENTS

9 lb. 2 oz. (4.1 kg) English 2-row pale malt (3°L)
1 lb. 2 oz. (0.51 kg) crystal malt (60°L)
2 lbs. (0.91 kg) flaked maize
5.25 AAU Target hops (60 min.) (0.53 oz./15 g of 10% alpha acids)
2.6 AAU Challenger hops (60 min.) (0.34 oz./10 g of 7.5% alpha acids)
0.83 AAU Northdown hops (15 min.) (0.1 oz./2.7 g of 8.5% alpha acids)

1.66 AAU Goldings hops (15 min.) (0.33 oz./9.4 g of 5% alpha acids)
¼ tsp. yeast nutrients
1 tsp. Irish moss
Wyeast 1968 (London ESB Ale), White Labs WLP002 (English Ale), or LalBrew London ESB yeast
¾ cup (150 g) dextrose (for priming)

STEP BY STEP

Mash the grains and maize in 15 quarts (14 L) of water at 154°F (68°C) for 60 minutes. Stir boiling water into the mash to boost the temperature to 168°F (76°C) and hold at that temperature for 5 minutes. Mash out, vorlauf, and then sparge at 170°F (77°C). Boil the wort for 90 minutes, adding hops to boil for the times indicated in ingredients list. Add the yeast nutrients and Irish moss with 15 minutes left in boil. Pitch the yeast and ferment at 70°F (21°C). With these yeast strains, you need to pitch an adequate amount of yeast and don't let the temperatures get too cool near the end of fermentation or the yeast may settle too soon. Rack to secondary when fermentation is complete. Bottle a few days later, when the beer falls clear.

EXTRACT WITH GRAINS OPTION: Substitute the English 2-row pale malt with 1.45 pounds (0.66 kg) Muntons light dried malt extract, 4 pounds (1.8 kg) Muntons light liquid malt extract (late addition), and 1 pound, 2 ounces (0.51 kg) English 2-row pale malt (3°L). Skip the flaked maize and add 1 pound, 5 ounces (0.60 kg) dextrose to the ingredients list.

In a large brewpot, heat 3.4 quarts (3.2 L) of water to 165°F (74°C). Add the crushed grains to the grain bag. Submerge the bag and let the grains steep around 154°F (68°C) for 45 minutes. While the grains steep, begin heating 2.25 gallons (8.5 L) of water in your brewpot. When the steep is over, remove 1.1 quarts (about 1.1 L) of water from the brewpot and add to the "grain tea" in the steeping pot. Place a colander over the brewpot and place the steeping bag in it. Pour the grain tea (with water added) through the grain bag. Heat the liquid in the brewpot to a boil, then stir in the dried malt extract, add the first charge of hops, and begin the 60-minute boil. With 15 minutes left in the boil, add the dextrose, remaining hops, yeast nutrients, and Irish moss. Turn off the heat and stir in the liquid malt extract. Stir well to dissolve the extract, then resume heating. (Keep the boil clock running while you stir.) At the end of the boil, cool the wort and transfer to your fermenter. Top up with filtered water to 5 gallons (19 L). Follow the remaining portion of the all-grain recipe.

GREAT DIVIDE BREWING COMPANY
YETI IMPERIAL STOUT

(5 gallons/19 L, all-grain) OG = 1.090 FG = 1.018 IBU = 75 SRM = 98 ABV = 9.3%

As big as the legend its namesake logo celebrates, Great Divide describes Yeti as "an onslaught of the senses." This beer bursts forth with roasty malt flavors and big, bold American hop aromas.

INGREDIENTS

15.25 lb. (6.9 kg) US 2-row pale malt
1 lb. (0.45 kg) crystal malt (120°L)
12 oz. (0.34 kg) chocolate malt
12 oz. (0.34 kg) black patent malt
10 oz. (0.28 kg) roasted barley
8 oz. (0.23 kg) flaked wheat
8 oz. (0.23 kg) flaked rye
14.3 AAU Chinook hops (60 min.)
  (1.1 oz./31 g at 13% alpha acids)

7.2 AAU Chinook hops (30 min.)
  (0.55 oz./16 g at 13% alpha acids)
5.3 AAU Centennial hops (15 min.)
  (0.50 oz./14 g at 10.5% alpha acids)
0.5 oz. (14 g) Centennial hops (5 min.)
Wyeast 1056 (American Ale), White Labs WLP001
  (California Ale), or SafAle US-05 yeast
³/₄ cup (150 g) dextrose (for priming)

STEP BY STEP

Mash the grains at 150°F (66°C). Mash out, vorlauf, and then sparge at 170°F (77°C). Boil for 60 minutes, adding the hops as indicated in the ingredients list. Following the boil, chill the wort rapidly to 68°F (20°C), transfer to a sanitized fermenter, aerate well, and pitch the yeast. Ferment at 70°F (21°C) until final gravity is reached. Bottle or keg as usual.

EXTRACT WITH GRAINS OPTION: Reduce the amount of 2-row pale malt to 2 pounds (0.91 kg). Add 9 pounds (4.1 kg) Muntons light liquid malt extract to the recipe. Heat 2.3 gallons (8.7 L) of water to 161°F (72°C). Submerge the grain bag(s) and partial mash at 150°F (66°C) for 30 to 45 minutes. **(Note: This is just over 6 pounds [2.7 kg] of grain, so you may need more than one grain bag depending on the size of the bag. Putting your brewpot in your oven on its lowest heat setting may help you maintain partial mash temperature.)** Remove the grains and rinse the grain bag(s) slowly with 1 gallon (3.8 L) of water at 170°F (77°C). Add water to the brewpot to make 4 gallons (15 L) of wort; stir in roughly two-thirds of the malt extract. Bring to a boil. Boil for 60 minutes, adding the hops at the times indicated in the ingredients list. Add the remaining malt extract with 15 minutes left in the boil. Chill the wort, transfer to your fermenter, and top up to 5 gallons (19 L) with filtered water. Follow the remaining portion of the all-grain recipe.

GREAT LAKES BREWING COMPANY
CHRISTMAS ALE

(5 gallons/19 L, all-grain) OG = 1.070 FG = 1.012 IBU = 30 SRM = 19 ABV = 7.5%

By Cleveland tradition, the annual release of Christmas Ale effectively marks the official start of the holiday season in the Forest City.

INGREDIENTS

15.25 lb. (6.9 kg) US 2-row pale malt
1 lb. (0.45 kg) crystal malt (120°L)
12 oz. (0.34 kg) chocolate malt
12 oz. (0.34 kg) black patent malt
10 oz. (0.28 kg) roasted barley
8 oz. (0.23 kg) flaked wheat
8 oz. (0.23 kg) flaked rye
14.3 AAU Chinook hops (60 min.)
  (1.1 oz./31 g at 13% alpha acids)

7.2 AAU Chinook hops (30 min.) (0.55 oz./16 g
  at 13% alpha acids)
5.3 AAU Centennial hops (15 min.) (0.50 oz./14 g
  at 10.5% alpha acids)
0.5 oz. (14 g) Centennial hops (5 min.)
Wyeast 1056 (American Ale), White Labs WLP001
  (California Ale), or SafAle US-05 yeast
³/₄ cup (150 g) dextrose (for priming)

STEP BY STEP

Mash the grains at 150°F (66°C). Mash out, vorlauf, and then sparge at 170°F (77°C). Boil for 60 minutes, adding the hops as indicated in the ingredients list. Following the boil, chill the wort rapidly to 68°F (20°C), transfer to a sanitized fermenter, aerate well, and pitch the yeast. Ferment at 70°F (21°C) until final gravity is reached. Bottle or keg as usual.

PARTIAL MASH OPTION: Reduce the amount of 2-row pale malt to 2 pounds (0.91 kg). Add 9 pounds (4.1 kg) Muntons light liquid malt extract to the recipe. Heat 2.3 gallons (8.7 L) of water to 161°F (72°C). Submerge the grain bag(s) and partial mash at 150°F (66°C) for 30 to 45 minutes. (Note: This is just over 6 pounds [2.7 kg] of grain, so you may need more than one grain bag depending on the size of the bag. Putting your brewpot in your oven on its lowest heat setting may help you maintain partial mash temperature.) Remove the grains and rinse the grain bag(s) slowly with 1 gallon (3.8 L) of water at 170°F (77°C). Add water to the brewpot to make 4 gallons (15 L) of wort; stir in roughly two-thirds of the malt extract. Bring to a boil. Boil for 60 minutes, adding the hops at the times indicated in the ingredients list. Add the remaining malt extract with 15 minutes left in the boil. Chill the wort, transfer to your fermenter, and top up to 5 gallons (19 L) with filtered water. Follow the remaining portion of the all-grain recipe.

GUINNESS & COMPANY
GUINNESS DRAUGHT

(5 gallons/19 L, all-grain) OG = 1.038 FG = 1.006 IBU = 45 SRM = 36 ABV = 4.2%

Guinness Draught, the kind found in widget cans or bottles, is an Irish dry stout. Guinness has a sharper roast character and more hop bitterness than Murphy's. The key to making a great clone is using roasted unmalted barley (or black barley) with a color rating around 500°L.

INGREDIENTS

5 lb. (2.3 kg) English 2-row pale ale malt
2.5 lb. (1.1 kg) flaked barley
1 lb. (0.45 kg) roasted barley (500°L)
12 AAU East Kent Golding hops (60 min.)
(2.4 oz./68 g at 5% alpha acids)

Wyeast 1084 (Irish Ale) or White Labs WLP004 (Irish Ale) yeast
3/4 cup (150 g) dextrose (if priming)

STEP BY STEP

Heat 2.66 gallons (10 L) of water to 161°F (72°C) and stir in crushed grains and flaked barley. Mash at 150°F (66°C) for 60 minutes. Stir boiling water into grain bed until temperature reaches 168°F (76°C) and rest for 5 minutes. Recirculate until wort is clear, then begin running wort off to kettle. Sparge with 170°F (77°C) water. Boil wort for 90 minutes, adding hops with 60 minutes left in boil. Cool wort and transfer to fermenter. Aerate wort and pitch yeast. Ferment at 72°F (22°C). Rack to secondary when fermentation is complete. Bottle a few days later, when beer falls clear. If beer is kegged, consider pushing with a nitrogen (beer gas) blend.

PARTIAL MASH OPTION: Reduce the 2-row pale malt in the all-grain recipe to 1.25 pounds (0.57 kg), reduce the flaked barley to 1.25 pounds (0.57 kg), and add 3.3 pounds (1.5 kg) Maris Otter liquid malt extract. Place crushed malt and barley in a steeping bag. In a large kitchen pot, heat 4.5 quarts (4.3 L) to 161°F (72°C) and submerge grain bag. Let grains mash for 45 minutes at around 150°F (66°C). While grains are steeping, begin heating 2.1 gallons (7.9 L) of water in your brewpot. Add in the crushed roasted barley and steep an additional 15 minutes. When steep is over, remove 1.25 quarts (1.2 L) of water from brewpot and add to the "grain tea" in steeping pot. Rinse grain bag with diluted grain tea and then bring to a boil. Stir in liquid malt extract off heat, and add hops at the beginning of the 60-minute boil. Top off to 5 gallons (19 L) at the end of the boil. Follow the remaining portion of the all-grain recipe.

TIPS FOR SUCCESS

To get that "Guinness tang," try this: After pitching the yeast to your stout, siphon 19 ounces of pitched wort to a sanitized 22-ounce bottle. Pitch bottle with a small amount of *Brettanomyces* and *Lactobacillus*. Cover bottle with aluminum foil and let ferment. When beer in bottle is done fermenting, pour it in a saucepan and heat to 160°F (71°C) for 15 minutes. Cool the beer and pour back in the bottle. Cap bottle and refrigerate. Add to stout when bottling or kegging.

HERETIC BREWING COMPANY
EVIL TWIN

(5 gallons/19 L, all-grain) OG = 1.064 FG = 1.014 IBU = 45 SRM = 17 ABV = 6.6%

Heretic describes Evil Twin as follows: "This blood-red ale may not be what you might expect from a malty and hoppy craft beer. Evil Twin has a rich malt character without being overly sweet. It has a huge hop character without being overly bitter."

INGREDIENTS

10.5 lb. (4.8 kg) British 2-row pale malt
1.6 lb. (0.74 kg) crystal malt (75°L)
14 oz. (0.4 kg) Munich malt
1 oz. (28 g) roasted barley
1 oz. (28 g) huskless black malt
4.3 AAU Columbus hops (60 min.) (0.31 oz./9 g at 14% alpha acids)
12.6 AAU Columbus hops (0 min.) (0.9 oz./25 g at 14% alpha acids)

9.9 AAU Citra hops (0 min.) (0.9 oz./25 g at 11% alpha acids)
0.8 oz. (23 g) Citra hops (dry hop, 7 days)
0.8 oz. (23 g) Columbus hops (dry hop, 7 days)
0.9 oz. (25 g) Citra hops (dry hop, 4 days)
White Labs WLP001 (California Ale), Wyeast 1056 (American Ale), or SafAle US-05 yeast
3/4 cup (150 g) dextrose (if priming)

STEP BY STEP

Mash the grains at 153°F (67°C) in 4.1 gallons (15.5 L) of water. Vorlauf until your runnings are clear. Sparge the grains with 2.8 gallons (10.6 L) and top up as necessary to obtain 6 gallons (23 L) of wort. Boil for 90 minutes. Add the first Columbus hops with 60 minutes left in the boil. Add the second Columbus hop addition and the first Citra addition at the end of the boil. After the boil, turn off the heat and begin a vigorous whirlpool in your kettle. Let the hot wort stand for 20 minutes, then chill the wort to slightly below fermentation temperature, about 65°F (18°C). Aerate the wort with pure oxygen or filtered air and pitch the yeast. Ferment at 67°F (19°C) for 7 days. Add the dry hops and raise to 72°F (22°C) for 3 more days. Add the second round of dry hops for an additional 4 days (7 total days of dry hopping). Once the beer reaches final gravity, bottle or keg the beer and carbonate to approximately 2.5 volumes. You may want to cold-crash the beer prior to packaging to 35°F (2°C) for 48 hours to improve the clarity.

TIPS FROM THE PROS

This beer's recipe has undergone several evolutions, from its origins as a homebrew recipe through its development as a commercial success for Heretic. This version came directly from the desk of Chief Heretic Jamil Zainasheff. The beer relies on extensive flameout and dry hopping additions to achieve its signature hop nose, with only a small bittering addition 30 minutes into its 90-minute boil. It even makes use of a second dry hopping round, receiving one last blast of Citra.

EXTRACT WITH GRAINS OPTION: Substitute the British 2-row pale malt with 8 pounds (3.6 kg) extra light liquid malt extract and omit the Munich malt. Bring 5.6 gallons (21 L) of water to approximately 162°F (72°C) and hold there. Steep the milled specialty grains in grain bags for 15 minutes. Remove the grain bags and let the bag drain fully. Add the liquid malt extract while stirring and stir until completely dissolved. Bring the wort to a boil. Boil for 90 minutes. Add the first Columbus hop charge with 60 minutes left in the boil. Add the second Columbus hop addition and the first Citra addition at the end of the boil. Follow the remaining portion of the all-grain recipe.

HILL FARMSTEAD BREWERY
EVERETT

(5 gallons/19 L, all-grain) OG = 1.088 FG = 1.030 IBU = 38 SRM = 55 ABV = 7.5%

Named for brewmaster Shaun Hill's grandfather's brother, this porter is brewed using American malted barley, English and German roasted malts, American hops, and the Hill Farmstead house ale yeast. It features deep dark flavors of coffee and chocolate.

INGREDIENTS

13.5 lb. (6.1 kg) 2-row pale malt
1.0 lb. (0.45 kg) dextrine malt
1.25 lb. (0.57 kg) caramalt (37°L)
0.5 lb. (0.23 kg) crystal malt (90°L)
1.0 lb. (0.45 kg) chocolate malt (300°L)
1.25 lb. (0.57 kg) roasted barley (500°L)

11 AAU Columbus hops (60 mins) (0.8 oz./23 g
  at 14% alpha acids)
Wyeast 1028 (London Ale), White Labs WLP013
  (London Ale), or LalBrew Nottingham yeast
¾ cup (150 g) dextrose (for priming)

STEP BY STEP

When crushing the grains, keep the dark roasted grains (crystal 90°L, chocolate malt, and roasted barley) separate from the other grains. Mash the 2-row pale malt, dextrine malt, and caramalt at 159°F (71°C) in 20 quarts (19 L) of water for 20 minutes, then mix in the darker grains. Hold for 5 minutes, then mash out, vorlauf, and sparge at 170°F (77°C). Boil for 60 minutes, adding the Columbus hops at the beginning of the boil. Pitch the yeast and ferment at 68°F (20°C) until the final gravity is reached, about 1 week. Allow the beer to condition for an additional week at 52°F (11°C) before packaging. Bottle or keg as usual.

EXTRACT WITH GRAINS OPTION: Replace the 2-row pale malt with 9.9 pounds (4.5 kg) light liquid malt extract. Crush the caramalt, crystal malt, chocolate malt, and roasted barley and place them in a muslin brewing bag. Steep the crushed specialty grains in 2 gallons (7.8 L) of water at 155°F (68°C) for 20 minutes. Rinse the grains with 3 quarts (2.8 L) of hot water and allow it to drip into the kettle for about 15 minutes. To prevent extracting harsh tannins from the grain husks, be sure not to squeeze the bag. Add the malt extract to the brewpot with the heat turned off and stir well to avoid scorching or boilovers. Top off the kettle to 6 gallons (23 L). Follow the remaining portion of the all-grain recipe.

TIPS FROM THE PROS

With the water he uses, Shaun Hill is able to start all of his Hill Farmstead beers with a fairly clean slate, carefully building the water profile for a beer like Everett. If you are attempting to brew this beer, he recommends that you make sure your water calculator can calculate for residual alkalinity (RA)–this is an important part of brewing with a large portion of dark grains when you want to keep the wort's pH in range. Using a water calculator such as BeerSmith or Bru'n Water will help. Shaun does not add any calcium chloride to Everett. If you haven't done so already, request a water report from your town or city (if you are using a municipal water source) or have your home water supply tested to find out what is in your water.

Everett

ROBUST AMERICAN PORTER

ICARUS BREWING COMPANY
PINEAPPLE HINDENBURG

(5 gallons/19 L, all-grain) OG = 1.064 FG = 1.016 IBU = 17.1 SRM = 6 ABV = 6.2%

This recipe was born out of a collaboration with the Ocean County Homebrewers club. The base recipe makes a very nice New England–style pale ale, but Icarus takes that and throws fresh habaneros into the kettle and conditions it on fresh pineapple that complements the BRU-1 hops.

INGREDIENTS

4.75 lb. (2.2 kg) Golden Promise pale ale malt
2.75 lb. (1.25 kg) spelt malt (wheat malt if unavailable)
2 lb. (0.9 kg) Vienna malt
1.75 lb. (0.8 kg) flaked oats
1 lb. (0.45 kg) wheat malt
1 lb. (0.45 kg) dextrin malt
0.16 lb. (73 g) acidulated malt
3.6 AAU CTZ hops (60 min.) (0.2 oz./5.7 g at 18% alpha acids)

3.9 AAU BRU-1 hops (0 min.) (0.3 oz./8.5 g at 13% alpha acids)
3 oz. (84 g) BRU-1 hops (dry hop)
3 oz. (84 g) Citra Cryo hops (dry hop)
0.75 oz. (21 g) chopped fresh habanero peppers (0 min.)
1 qt. (0.9 L) pineapple, puréed
White Labs WLP066 (London Fog), Wyeast 1318 (London Ale III), Imperial Yeast A38 (Juice), or LalBrew Verdant IPA yeast
$^2/_3$ cup (130 g) dextrose (for priming)

STEP BY STEP

This is a single-infusion mash with a soft water profile that favors chlorides over sulfates with a dash of Epsom salt (150 ppm chloride, 40 ppm sulfate, 15 ppm magnesium). Strike in utilizing 1.3 quarts of water per pound of grain (2.7 L/kg) to achieve a stable mash temperature around 152°F (67°C). Hold for 60 minutes or until conversion is complete. Begin the lauter process by either raising to mash out temperature or recirculating. Sparge with enough 168°F (76°C) water to collect 6.5 gallons (24.6 L) of wort. Add the CTZ hops at the beginning of the boil and then boil 60 minutes.

Turn off heat and add flameout hops and habanero peppers, then begin a whirlpool. Allow wort to settle for 15 minutes before chilling to 66°F (19°C). Aerate the wort with filtered air (you don't want to overoxygenate). Pitch yeast and ferment at 68°F (20°C). Add the dry hops 1 day after signs of fermentation. After 5 days, transfer onto pineapple in a well-purged keg or secondary vessel for several days. If serving on draft, you may want to transfer to another purged keg. Carbonate the beer to 2.3 volumes CO_2.

PARTIAL MASH OPTION: Eliminate the pale ale, spelt, and acidulated malts and reduce the dextrin malt to 0.5 pound (0.23 kg). Replace with 3.3 pounds (1.5 kg) Maris Otter liquid malt extract and 2.5 pounds (1.1 kg) wheat dried malt extract and ½ teaspoon, 88 percent lactic acid. In a large muslin bag, add the crushed Vienna and dextrin malts as well as the oats. Heat 5.5 quarts (5.2 L) of water to 164°F (73°C) and submerge the grains in the water. Mash at around 152°F (67°C) for 45–60 minutes. Remove the grains and place in a large colander. Slowly wash the grains with 1 gallon (4 L) of hot water. Bring volume up to 6.25 gallons (23.7 L) and stir in the malt extracts and lactic acid. Once the extracts are fully dissolved, bring the wort up to a boil. Follow the remainder of the all-grain recipe.

JACK'S ABBY BREWING
HOPONIUS UNION

(5 gallon/19 L, all-grain) OG = 1.063 FG = 1.014 IBU = 65 SRM = 6 ABV = 6.7%

Hoponius Union is a lager that features a blend of popular American hops, creating a huge tropical fruit and citrusy hop aroma. A dry finish accentuates the pleasant bitterness and hop profile.

INGREDIENTS

10.5 lb. (4.5 kg) 2-row pale malt
1.6 lb. (0.73 kg) Munich malt (9°L)
9 oz. (255 g) Weyermann Carabelge malt (12°L)
9 oz. (255 g) spelt berries
3.3 AAU Magnum pellet hops (60 min.)
  (0.25 oz./7 g at 13% alpha acids)
16.5 AAU Centennial pellet hops (5 min.)
  (1.5 oz./43 g at 11% alpha acids)
22 AAU Centennial pellet hops (0 min.)
  (2 oz./57 g at 11% alpha acids)

26 AAU Citra pellet hops (0 min.) (2 oz./57 g
  at 13% alpha acids)
1.5 oz. (43 g) Centennial pellet hops (dry hop)
1.5 oz. (43 g) Citra pellet hops (dry hop)
White Labs WLP830 (German Lager), Wyeast 2124
  (Bohemian Lager), SafLager W-34/70, White
  Labs WLP029 (German Ale/Kölsch), or Wyeast 2565
  (Kölsch) yeast
3/4 cup (150 g) dextrose (if priming)

STEP BY STEP

Mash the grains at 148°F (64°C) for 90 minutes. Mash out, vorlauf, and then sparge at 168°F (76°C). Boil for 60 minutes, adding hops at the times indicated. At the end of the boil, remove from heat and add the 0-minute hops. Begin a whirlpool and let settle for about 20 minutes. At the end of the 20 minutes, cool the wort to yeast pitching temperature, 65°F (18°C) if using a Kölsch strain and 50°F (10°C) if using the lager strain. Pitch the yeast and ferment at 65°F (18°C) for the Kölsch strain and 50 to 55°F (10 to 12°C) for the lager strain. Dry hop at the same temperature and then lager at 32°F (0°C) for 3 weeks. Bottle or keg as usual.

PARTIAL MASH OPTION: Substitute 7 pounds (3.2 kg) golden light liquid malt extract for the 2-row malt and scale the Weyermann Carabelge malt down to 8 ounces (227 g). Place the crushed grains in a large nylon bag. Mix grains with 1 gallon of 158°F (70°C) water to achieve a stable mash temperature at 148°F (64°C) and hold for 60 minutes. Raise temperature of the mash by either direct heat or adding boiling water to 168°F (76°C) and hold for 5 minutes. Remove the grain bag and place in a colander. Slowly pour 1 gallon (3.8 L) of hot water over the grains. Add the liquid malt extract off heat, then add water in your brewpot until you have about 6 gallons (23 L) of wort in your kettle, then bring up to a boil. Boil for 60 minutes. Follow remaining portion of all-grain recipe.

JACKIE O'S BREWERY
BRICK KILN

(5 gallons/19 L, all-grain) OG = 1.101 FG = 1.024 IBU= 35 SRM = 28 ABV = 10.2%

Brick Kiln from Jackie O's Brewery (Athens, Ohio) is a barleywine brewed in the English tradition and features rich caramel and raisin notes, with a light herbal presence from noble hops.

INGREDIENTS

10 lb. (4.5 kg) 2-row malt
6 lb. (2.7 kg) Maris Otter malt
1.6 lb. (0.68 kg) Munich malt (10°L)
1.2 lb. (0.54 kg) Special B malt
1.2 lb. (0.54 kg) caramel malt (45°L)
1 lb. (0.45 kg) white unmalted wheat
4 oz. (113 g) chocolate malt
½ tsp. yeast nutrients (10 min.)
3.5 AAU Magnum hops (90 min.)
  (0.25 oz./7 g at 14% alpha acids)

3.5 AAU Magnum hops (45 min.)
  (0.25 oz./7 g at 14% alpha acids)
2.4 AAU East Kent Golding hops (30 min.)
  (0.5 oz./14 g at 4.75% alpha acids)
2.4 AAU East Kent Golding hops (15 min.)
  (0.5 oz./14 g at 4.75% alpha acids)
1 Bourbon barrel or 1–2 oz. (28-57 g) medium
  toasted oak chips, soaked in Bourbon
Wyeast 1056 (American Ale), White Labs WLP001
  (California Ale), or SafAle US-05 yeast
⅔ cup (130 g) dextrose (for priming)

STEP BY STEP

This is a single-infusion mash, targeting about 1.15 quarts/pound (2.4 L/kg) water-to-grain mash ratio, about 24.4 quarts (23.1 L). Mash at 152°F (67°C) for 45–60 minutes. Sparge with approximately 17 quarts (16.1 L) to collect approximately 7.5 gallons (28 L). Concentrate your wort by boiling for at least 2 hours, with the goal to end the boil with 5.25 gallons (20 L) in the kettle. Add hops starting with about 90 minutes left in the boil.

When boil is complete, cool, and then oxygenate for at least 2 minutes. Pitch yeast that has been built up in a yeast starter or two sachets of dried yeast. Ferment at 68°F (20°C) for 2 days, raise to 70°F (21°C) on day 3, then free rise on day 4. Keep in primary longer than a normal beer to allow the yeast to finish fermentation and clean up by-products, about 2–4 weeks. If you are not barrel-aging, condition for 3–6 months. If you are Bourbon-barrel aging, you will need to taste periodically to find the proper aging time. Jackie O's ages theirs for a better part of a year in Bourbon barrels, but you need to determine when it is right for barrel and beer. Carbonate to your personal preference, but 2.2 volumes of CO_2 seems appropriate.

PARTIAL MASH OPTION: Replace the 2-row malt with 6 pounds (2.7 kg) extra light dried malt extract and the Maris Otter malt with 3.3 pounds (1.5 kg) Maris Otter liquid malt extract. Some of the grains need to be mashed (Munich and the white wheat) while the others just need to be steeped. Use two muslin bags and put the Munich and white wheat in one bag, the remaining grains in the other. Put 6.6 quarts (6.2 L) of water in your pot and raise the temperature to 160°F (71°C). Submerge the Munich and white wheat in the water, making sure the grains get thoroughly mixed in. Temperature should stabilize at 152°F (67°C) and hold there for 30 minutes. Add the second bag with the steeping grains and hold for 15 minutes. When done, place both grain bags in a colander and slowly pour 1.5 gallons (5.7 L) of water to wash the grains. Top off your pot to 6 gallons (23 L) total. Raise to a boil, take the pot off the heat, and add the malt extracts while stirring thoroughly. Boil for 60 minutes. Follow the remainder of the all-grain recipe.

JOLLY PUMPKIN ARTISAN ALES
LA ROJA

(5 gallons/19 L, all-grain) OG = 1.062 FG = will vary IBU = 25 SRM = 21 ABV = around 7%

La Roja is one of Jolly Pumpkin's first and signature beers. Founder and brewmaster Ron Jeffries says that it is loosely based in the Flanders sour red tradition.

INGREDIENTS

8 lb. 5 oz. (3.8 kg) blend of Pilsner and 2-row pale malts
1 lb. (0.45 kg) malted wheat
1 lb. 4 oz. (0.57 kg) Munich malt (10°L)
13 oz. (0.37 kg) crystal malt (120°L)
0.5 oz. (14 g) black malt
1 lb. 2 oz. (0.51 kg) dextrose (added to kettle)

4 AAU Hallertau hops or other noble hop (60 min)
 (1 oz./28 g of 4% alpha acids)
4 AAU Hallertau hops or other noble hop (30 min)
 (1 oz./28 g of 4% alpha acids)
Wyeast 3763 (Roeselare Ale) blend
1 cup (200 g) dextrose (for priming)

STEP BY STEP

Mash the grains at 154°F (68°C) and hold for 60 minutes. Mash out, vorlauf, and then sparge at 170°F (77°C). Boil for 60 minutes, adding the hops at the beginning of the boil and again with 30 minutes remaining. When the boil is finished, chill the wort rapidly to 65°F (18°C). Ferment at 68°F (20°C), then rack to barrel or into a fermenter with oak cubes (or other alternative oak) for aging. Wyeast 3763 (Roeselare Ale) is a blend of lambic cultures, including a Belgian-style ale strain, a sherry strain, two *Brettanomyces* strains, a *Lactobacillus* culture, and a *Pediococcus* culture. Wyeast recommends aging beers brewed with this yeast for up to 18 months to get a full flavor profile and for the acidity to develop. Bottle or keg as usual; heavy-duty bottles are recommended.

EXTRACT WITH GRAINS OPTION: Omit the 2-row pale and Pilsner malts. Add 1 pound (0.45 kg) Muntons light dried malt extract and 4 pounds 14 ounces (2.2 kg) Muntons light liquid malt extract. Steep the grains in 4.6 quarts (4.4 L) of water at 154°F (68°C) for 45 minutes. Rinse with 2.3 quarts (about 2.2 L) of water at 170°F (77°C). Add water to make 3 gallons (11 L), add the dried malt extract, and bring to a boil. Boil for 60 minutes, stirring in the liquid malt extract for the final 15 minutes of the boil. Chill the wort, transfer to your fermenter, and top up with filtered water to 5 gallons (19 L). Follow the remaining portion of the all-grain recipe.

THE KERNEL BREWERY
EXPORT INDIA PORTER

(5 gallons/19 L, all-grain) OG = 1.060 FG = 1.015 IBU = 48 SRM = 40 ABV = 6%

This recipe is based on some of the Barclay Perkins (1855) and Whitbread (1856) porters that were sent out to India nearly two hundred years ago. Of course, elements of the ingredients, equipment, and processes are different, and tastes have also changed over time, so in keeping with the Kernel's philosophy, they have made a beer that contemporary beer drinkers want to drink, rather than a blindly faithful copy of a nineteenth-century recipe.

INGREDIENTS

9.6 lb. (4.3 kg) Maris Otter pale ale malt
0.9 lb. (0.4 kg) brown malt (38°L)
0.9 lb. (0.4 kg) chocolate malt (330°L)
0.9 lb. (0.4 kg) crystal malt (60°L)
0.4 lb. (0.2 kg) black malt (500°L)
5.6 AAU Bramling Cross hops (first wort hop)
  (0.7 oz./20 g at 8% alpha acids)
5.6 AAU Bramling Cross hops (15 min.)
  (0.7 oz./20 g at 8% alpha acids)

AAU Bramling Cross hops (10 min.)
  (1 oz./30 g at 8% alpha acids)
11.2 AAU Bramling Cross hops (5 min.)
  (1.4 oz./40 g at 8% alpha acids)
2.8 oz. (80 g) Bramling Cross hops (dry hop)
½ Whirlfloc tablet (10 min.)
White Labs WLP013 (London Ale), Wyeast 1028 (London Ale), SafAle S-04, or LalBrew Nottingham yeast
½ cup (100 g) cane sugar (if priming)

STEP BY STEP

Mill the grains and dough-in targeting a mash of around 1.3 quarts of strike water to 1 pound of grain (2.7 L/kg) and a temperature of 154°F (68°C). Hold the mash at 154°F (68°C) until enzymatic conversion is complete. Sparge slowly with 171°F (77°C), collecting wort until the preboil kettle volume is 6 gallons (23 L). Add the first wort hops during the sparging process. Boil for 60 minutes, adding hops and Whirlfloc at times indicated.

Chill the wort to 68°F (20°C) and aerate thoroughly. Pitch the yeast. Ferment at 68°F (20°C) until fermentation is complete. Dry hop 3 days before bottling or kegging. Carbonate the beer to around 2.4 volumes of CO_2. Condition at 59–68°F (15–20°C), allowing time for the beer to carbonate fully.

EXTRACT WITH GRAINS OPTION: Replace the pale ale malt with 6.6 pounds (3 kg) Maris Otter liquid malt extract. Steep the crushed grains in 6 gallons (23 L) of 154°F (68°C) water for 30 minutes. Remove steeping bag and add liquid malt extract with the heat source off. Stir until completely dissolved. Add the first wort hops and bring wort to a boil. Total boil time is 60 minutes. Follow the remaining portion of the all-grain recipe.

TIPS FOR SUCCESS

Chocolate malt and crystal malt has replaced some of the black malt, which can sometimes produce burnt or astringent flavors. This recipe calls for lots of late hops and dry hopping and the Kernel Brewery experiments with a number of different varieties. Bramling Cross (used here) gives a traditional British character, but Columbus can also work well if you want more of a New World character.

KNOTTED ROOT BREWING COMPANY
PERPETUALLY UNIMPRESSED

(5 gallon/19 L, all-grain)   OG = 1.076   FG = 1.015   IBU = 80   SRM = 7   ABV = 8.1%

INGREDIENTS

6 lb. (2.7 kg) Pilsner malt
5.5 lb. (2.5 kg) Simpsons Golden Promise pale ale malt
1.6 lb. (0.71 kg) flaked oats
1 lb. (0.45 kg) Rahr white wheat malt
1 lb. (0.45 kg) Golden Naked Oats malt
7 oz. (200 g) honey malt
4 oz. (113 g) acidulated malt

6 oz. (170 g) Citra hops (hop stand)
10 oz. (283 g) Nelson Sauvin hops (dry hop #1)
10 oz. (283 g) Nelson Sauvin hops (dry hop #2)
Wyeast 1318 (London Ale III), Omega OYL-052 (Conan),
   or LalBrew New England yeast
3/4 cup (150 g) dextrose (if priming)

STEP BY STEP

Mill the grains, then mix with 4.9 gallons (18.5 L) of 167°F (75°C) strike water to achieve a single-infusion rest temperature of 152°F (67°C). At this time, add 50 ppm of gypsum. Hold at this temperature for 60 minutes. Mash out to 170°F (77°C) if desired. Vorlauf until your runnings are clear before directing them to your boil kettle. Batch or fly sparge the mash to obtain 6.5 gallons (25 L) of wort. Preboil pH should 5.2 to 5.4. Boil for 90 minutes. After the boil, add 150 ppm of calcium chloride, cool the wort to approximately 190°F (88°C), and add the Citra hops. Whirlpool for 40 minutes before further chilling the wort to 68°F (20°C). Pitch yeast. Maintain rough fermentation temperature but allow for a free rise to 72°F (22°C) by the end of primary fermentation for this beer.

Rack the beer off the trub or drop the cone at ambient temperature. Add the dry hops sequentially as indicated and let them extract for 1 day each. During this time shake the carboy or degas with CO_2 to increase the rate of hop oil extraction. Cold-crash for 24 hours before bottling or kegging the beer. Carbonate to approximately 2.3–2.4 volumes.

TIPS FOR SUCCESS

With such a fruity, hop-forward beer, you could correctly assume that water chemistry is quite important. For the all-grain brewer, shoot for a 3:1 ratio of chloride-to-sulfate prior to pitching. Unfortunately, for the extract brewer, you're at the whims of the maltster. If you're feeling adventurous, add 50–100 ppm of calcium chloride to tilt the balance to chlorides. Err on the side of caution as too much may leave a "chemical" taste on the tongue. The other major consideration to concern yourself with is the yeast pitch rate. With an OG = 1.076, you will be best off using a fresh starter.

EXTRACT WITH GRAINS OPTION: Replace the Pilsner, pale ale, wheat, and acidulated malts with 3.3 pounds (1.5 kg) Pilsen, 3 pounds (1.36 kg) pale ale, and ¾ pound (0.34 kg) wheat dried malt extracts and ½ teaspoon lactic acid. Bring 5.5 gallons (21 L) water to roughly 150°F (66°C). Steep both types of oats and the honey malt for 15 minutes before removing and draining. Add all the types of DME, with stirring, before heating to a boil. Add the lactic acid, then boil for 15 minutes. Follow the remainder of the all-grain recipe instructions for post-boil and fermentation directions.

CHAPTER 3

LAUNCH PAD BREWERY
CAPE CANAVERAL KEY LIME SAISON

(5 gallons/19 L, all-grain) OG = 1.067 FG = 1.013 IBU = 24 SRM = 7 ABV = 7.5%

Lactose isn't the only twist on this saison from Launch Pad Brewery (Aurora, Colorado), which also features lime peels, juice, and leaves, as well as vanilla beans to create a beer reminiscent of a key lime pie.

INGREDIENTS

7 lb. (3.2 kg) Belgian Pilsner malt
3.75 lb. (1.7 kg) Maris Otter pale malt
10 oz. (0.22 kg) crystal malt (45°L)
6 oz. (0.17 kg) flaked oats
1 lb. 2 oz. (0.51 kg) lactose sugar (10 min.)
4.5 AAU Sorachi Ace hops (60 min.)
  (0.4 oz./11 g at 11.4% alpha acids)

4 AAU Sorachi Ace hops (5 min.) (0.35 oz./10 g
  at 11.4% alpha acids)
2 lb. (0.91 kg) fresh key limes (peeled and juiced)
0.1 oz. (3 g) Makrut (Kaffir) lime leaves
2 Madagascar Bourbon vanilla beans
Inland Island INIS-291 (Saison: Farmhouse), Wyeast 3724
  (Belgian Saison), or LalBrew Belle Saison yeast
1 cup (200 g) dextrose (if priming)

STEP BY STEP

Peel fresh key limes and juice. According to owner David Levesque of Launch Pad Brewery, he suggests 2 ounces of key lime peel, chopped up in a good processor to smaller pieces to provide more surface area. Mash in with 2.5 gallons (9.5 L) of water, aiming for 150°F (66°C) strike temperature. Hold 1.5 hours or until converted. Raise to 168°F (76°C) for mashout. Hold 10 minutes. Sparge with enough water to collect about 6.5 gallons (24.6 L) of wort. Boil for 75 minutes total. Add hops as indicated and key lime peel and lactose with 10 minutes remaining in the boil.

Ferment at 75°F (24°C). When done, rack to secondary and add key lime juice and two Madagascar Bourbon vanilla beans (Launch Pad slices theirs lengthwise, scrapes them out, and soaks in about 4 ounces/118 mL vodka). Condition for 2 weeks on the vanilla and rack to keg or bottles.

PARTIAL MASH OPTION: Reduce the Maris Otter malt to 2 pounds (0.91 kg) and replace all the Pilsner malt with 3.5 pounds (1.59 kg) Pilsen dried malt extract and 1.25 pounds (0.57 kg) Muntons dried malt extract. Peel fresh key limes and juice. Place crushed malt in a muslin bag. Mash the grains in 1 gallon (4 L) of water at 150°F (66°C) for 60 minutes or until converted. Remove grain bag and wash with 2 quarts (2 L) of hot water. Top off the kettle to 6 gallons (23 L) and raise to a boil. Once boiling, remove kettle from heat and stir in dried malt extracts. Boil for 60 minutes, adding hops as indicated and key lime peel and lactose with 10 minutes remaining.

Ferment at 75°F (24°C). When done, rack to secondary and add key lime juice and two Madagascar Bourbon vanilla beans (Launch Pad slices theirs lengthwise, scrapes them out, and soaks in about 4 ounces/118 mL vodka). Condition for 2 weeks on the vanilla and rack to keg or bottles.

#CRAFTBEERISOURROCKETFUEL

LAUNCH PA
BREWER

Cape Canaveral
Key Lime Saison
Blonde-Lime-Graham
7 % ABV

LAWSON'S FINEST LIQUIDS
DOUBLE SUNSHINE

(5 gallons/19 L, all-grain) OG = 1.074 FG = 1.013 IBU = 100+ SRM = 6 ABV = 8%

Double Sunshine is a sought-after "Vermont-style" Double IPA. It's packed with juicy tropical fruit flavors and bright herbal aromas thanks to the abundance of US-grown Citra hops.

INGREDIENTS

- 9.5 lb. (4.3 kg) 2-row pale malt
- 2.5 lb. (1.1 kg) Vienna-style malt
- 1 lb. (0.45 kg) flaked oats
- 12 oz. (0.34 kg) carapilsen malt (7–9°L)
- 6 oz. (0.17 kg) caramunich-type malt (20–30°L)
- 1 lb. (0.45 kg) dextrose (10 min.)
- 10.5 AAU Columbus hops (60 min.)
  (0.75 oz./21 g at 14% alpha acids)
- 12.5 AAU Citra hops (20 min.)
  (1 oz./21 g at 12.5% alpha acids)

- 37.5 AAU Citra hops (5 min.)
  (3 oz./84 g at 12.5% alpha acids)
- 37.5 AAU Citra hops (knockout)
  (3 oz./84 g at 12.5% alpha acids)
- 3 oz. (84 g) Citra hops (dry hop)
- SafAle US-05, LalBrew BRY-97, Wyeast 1056 (American Ale), or White Labs WLP001 (California Ale) yeast
- 3/4 cup (150 g) dextrose (for priming)

STEP BY STEP

A few days before brew day, make a yeast starter. You will need 254 billion healthy yeast cells. Mash at 152°F (67°C) for 45 minutes. Mash out, vorlauf, and then sparge at 170°F (77°C). Boil for 60 minutes, adding the hops as instructed and the dextrose with 10 minutes left in the boil. After the boil is complete, begin a whirlpool in the kettle and let the knockout hops rest in the hot wort for at least 30 minutes before chilling. Chill the wort, pitch the yeast, and ferment at 68°F (18°C) for 1 week. Cool to 55°F (13°C) to settle the yeast. Dump the yeast from the bottom of fermenter or rack to a clean, sanitized vessel. Add the dry hops and let the beer sit for an additional 4 to 7 days at 55–57°F (13–14°C). Bottle or keg as usual.

EXTRACT WITH GRAINS OPTION: Substitute the 2-row pale malt in the all-grain recipe with 6.6 pounds (3 kg) light liquid malt extract, skip the carapilsen malt, and boost the dextrose up to 1.5 pounds (0.68 kg). Mix the crushed Vienna-style malt, flaked oats, and caramunich-type malts into 2 gallons (7.6 L) of water to achieve a temperature of 152°F (67°C), then hold at this temperature for 45 minutes. Rinse the grains with 2.5 quarts (2.4 L) of hot water, add the liquid extract, and bring to a boil. Top off the kettle to 6.5 gallons (25 L). Boil for 60 minutes. Follow the remaining portion of the all-grain recipe.

TIPS FROM THE PROS

Sean Lawson advises homebrewers to "determine if you have hard or soft water. If you have hard water, then cut by at least half with distilled or reverse osmosis (RO). With soft water, a basic guideline for IPAs would be to add equal parts gypsum and calcium chloride to bring total calcium content over 50 ppm." If you homebrew with a water source that is chlorinated, one Campden tablet in 20 gallons (76 L) of water (left overnight) will rid your water of chlorine compounds that can lead to off-flavors in your beer.

LEFT HAND BREWING COMPANY
OKTOBERFEST

(5 gallons/19 L, all-grain) OG = 1.061 FG = 1.010 IBU = 25 SRM = 9 ABV = 6.6%

This beer is crystal clear with copper hues. Biscuit and bread crust notes lead, supported by noble hop character that provides a dry and spicy finish.

INGREDIENTS

7.25 lb. (3.3 kg) Munich malt (10°L)
5.7 lb. (2.6 kg) Pilsner malt
4.5 AAU Apollo hops (60 min.) (0.25 oz./7.1 g
  at 18% alpha acids)
2.5 AAU Mt. Hood hops (25 min.) (0.5 oz./14.2 g
  at 5% alpha acids)

1.75 AAU Mt. Hood hops (5 min.) (0.35 oz./9.9 g
  at 5% alpha acids)
1 Whirlfloc tablet (25 min.)
Wyeast 2352 (Munich Lager II), White Labs WLP860
  (Munich Helles Lager), or SafLager S-189 yeast
$^3/_4$ cup (150 g) dextrose (if priming)

STEP BY STEP

Mill the grains, then mix with 5 gallons (19 L) of approximately 131°F (55°C) water to achieve a 122°F (50°C) mash temperature. Let the mash rest for 10 minutes. During the mash rest, check the pH. The target is 5.4 pH; add lactic acid to achieve the target. Once the pH is within range, raise the temperature of the mash to 147°F (64°C) and hold for 30 minutes. Next, raise the temperature of the mash to 154°F (68°C) and hold for 30 minutes. Transfer 40 percent of the mash to a separate vessel and bring to a boil. Boil for 15 minutes. The aroma will be marvelous. Transfer boiled mash back to lauter tun.

Vorlauf for 15 minutes or until wort is clear. Run off the wort to the kettle, then begin to sparge once you have 1–2 gallons (3.8–7.6 L) wort in the kettle. Due to the decoction, keep the mash bed fairly wet to avoid a stuck bed. Sparge four gallons (15.1 L) water at 172°F (78°C). Lauter until you have 6.5 gallons (24.6 L) of wort. Boil wort for 60 minutes, adding hops and Whirlfloc at indicated times. Once the boil is complete, whirlpool the wort and let rest for 15 minutes.

TIPS FOR SUCCESS

The all-grain version will provide a more "true-to-the-original" representation of the beer because, according to Head Brewer Adam Lawrence, "The decoction is everything in this beer." But those who are not yet ready to take on all-grain or decoctions will find the extract recipe results in a suitable and tasty beer.

Chill the wort to 50°F (10°C). Add a healthy yeast pitch to the fermenter and aerate with pure oxygen (unless using dry yeast strain). Ferment at 52°F (11°C) until the beer is below 1.030, then increase temperature slightly to 57°F (14°C) for a diacetyl rest. Leave at that temperature until beer has been terminal and stable for 3 days. Drop temperature to 32°F (0°C), dump the yeast after 10 days and lager for at least 30 more days. Bottle or keg and carbonate to 2.6 v/v.

EXTRACT-ONLY OPTION: Swap out the Munich and Pilsner malts with 3.8 pounds (1.7 kg) Munich dried malt extract and 3 pounds (1.36 kg) Pilsen dried malt extract. Start by heating 3 gallons (11.4 L) of water in your kettle. Bring to, or at least near, a boil. Remove from heat and stir in the malt extract slowly to avoid clumping. Once all extract is in and dissolved, return to boil (for 60 minutes). Once the boil is complete, whirlpool the wort and let rest for 15 minutes.

Chill the wort to 50°F (10°C). Top your fermenter with preboiled and chilled water also at 50°F (10°C) to yield a total of 5 gallons (19 L). Follow the remaining portion of the all-grain recipe.

THE LOST ABBEY BREWING COMPANY
TEN COMMANDMENTS

(5 gallons/19 L, all-grain) OG = 1.089 FG = 1.006 IBU = 34 SRM = 33 ABV = 11%

Ten Commandments is a multilayered beer that changes like a chameleon as it warms. In the nose, you'll find quite a bit of caramel candy, honey, raisin, chocolate with supportive herby rosemary, peppery spice, and restrained funky notes from the *Brettanomyces*. The flavor follows the nose for the most part and starts out a bit on the sweet side, which tails off in a long, drying finish. Pepper, funk, caramel, and raisin linger on the palate.

INGREDIENTS

10 lb. (4.54 kg) 2-row pale malt
1.5 lb. (0.68 kg) crystal wheat malt (55°L)
0.75 lb. (0.34 kg) Special B malt
0.75 lb. (0.34 kg) melanoidin malt
0.75 lb. (0.34 kg) flaked barley
0.4 lb. (0.18 kg) Carafa II malt
1.4 lb. (0.64 kg) dextrose
1.4 lb. (0.64 kg) honey
6.75 AAU Amarillo hops (90 min.)
  (0.75 oz./21 g at 9% alpha acids)

3.25 AAU Magnum hops (45 min.) (0.25 oz./7 g
  at 13% alpha acids)
5 oz. (142 g) blackened raisins (see Tips for Success)
0.5 oz. (14 g) sweet orange peel
0.026 oz (0.75 g) fresh rosemary
WLP565 (Belgian Saison), Wyeast 3724
  (Belgian Saison), or LalBrew Belle Saison yeast
WLP650 (Brettanomyces bruxellensis) or Wyeast 5112
  (Brettanomyces bruxellensis) yeast
1 cup (200 g) dextrose (if priming)

STEP BY STEP

Mill the grains, then mix with 4.4 gallons (16.7 L) of 166°F (74°C) strike water to achieve a single-infusion rest temperature of 152°F (67°C). Hold at this temperature for 60 minutes. Mash out to 170°F (77°C). Vorlauf until your runnings are clear before directing them to your boil kettle. Batch or fly sparge the mash to obtain 7 gallons (26.5 L) of wort. Boil for 90 minutes, adding hops at the times indicated. At 15 minutes left in the boil, you can add Irish moss or Whirlfloc as kettle fining agents.

After the boil, add the dextrose, honey, raisins, orange peel, and rosemary. Whirlpool for 15–20 minutes before chilling the wort to slightly below fermentation temperature. Pitch saison yeast. Start fermentation around 75°F (24°C) and ramp up as it goes. Ferment to completion, which may require a bit of patience and time. Bottle or keg the beer and carbonate to approximately three volumes using Brett Brux yeast.

TIPS FOR SUCCESS

Co-founder and Chief Operating Owner Tomme Arthur recommends patience with this recipe as all the flavors in the beer take months to meld together.

EXTRACT WITH GRAINS OPTION: Replace all the 2-row pale malt with 5.5 pounds (4.5 kg) extra light dried malt extract. Bring 2.5 gallons (9.5 L) of water to roughly 152°F (67°C). Steep all the milled malt in a nylon bag for 30 minutes, then remove. Allow the bag to drain back into the kettle. Add enough water to bring the total volume to 6.5 gallons (24.6 L). Add the dried malt extract, stir, and finally heat to a boil. Follow the remaining portion of the all-grain recipe.

NEW BELGIUM BREWING COMPANY
FAT TIRE AMBER ALE

(5 gallons/19 L, all-grain) OG = 1.050   FG = 1.013   IBU = 20   SRM = 14   ABV = 4.8%

New Belgium Brewing Company's flagship beer is named for a bike trip the brewery's co-founder took through Europe. This clear, bright, medium-bodied amber ale features sweet, biscuity, caramel malts and shows off with subtle notes of fennel and green apple.

INGREDIENTS

8 lb. 10 oz. (3.9 kg) 2-row pale malt
0.5 lb. (0.23 kg) Munich malt
0.5 lb. (0.23 kg) Carapils malt
0.5 lb. (0.23 kg) crystal malt (20°L)
6 oz. (168 g) biscuit malt
1 oz. (28 g) chocolate malt
4.3 AAU Willamette hops (60 min.) (0.86 oz./24 g at 5% alpha acid)

2 AAU Fuggle hops (20 min.) (0.40 oz./11 g at 5% alpha acid)
2 AAUs Fuggle pellet hops (0 min.) (0.40 oz./11 g at 5% alpha acid)
1 tsp. Irish moss (15 min.)
Wyeast 1272 (American Ale II), White Labs WLP051 (California Ale V), or Mangrove Jack's M36 (Liberty Bell) yeast
³/₄ cup (150 g) dextrose (for priming)

STEP BY STEP

Mash the grains at 154°F (68°C) in 13 quarts (12 L) of water for 45 minutes. Mash out, vorlauf, and then sparge at 170°F (77°C) to collect 6 gallons (23 L) of wort. Add 0.5 gallon (1.9 L) of water and boil for 60 minutes, adding the hops at the times indicated in the ingredients list. Add the Irish moss with 15 minutes left in the boil. Pitch the yeast and ferment at 68°F (20°C) until final gravity is reached (7 to 10 days). Bottle or keg with dextrose. (Try lowering the amount of priming sugar to mimic the low carbonation level of Fat Tire.) Lay the beer down for at least a few months to mellow and mature for best results.

EXTRACT WITH GRAINS OPTION: Omit the 2-row pale malt and instead use 2 pounds 3 ounces (1.0 kg) Coopers light dried malt extract and 3.3 pounds (1.5 kg) Coopers light liquid malt extract (late addition). Place the crushed malts in a nylon steeping bag and steep in 3 quarts (2.8 L) of water at 154°F (68°C) for 30 minutes. Rinse the grains with 1.5 quarts (about 1.5 L) of water at 170°F (77°C). Add water to make 3 gallons (11 L), stir in the dried malt extract, and bring to a boil. Boil for 60 minutes, adding the hops at the times indicated in the ingredients list. Add the liquid malt extract and Irish moss with 15 minutes left in the boil. Chill the wort, transfer to your fermenter, and top up with filtered water to 5 gallons (19 L). Follow the remaining portion of the all-grain recipe.

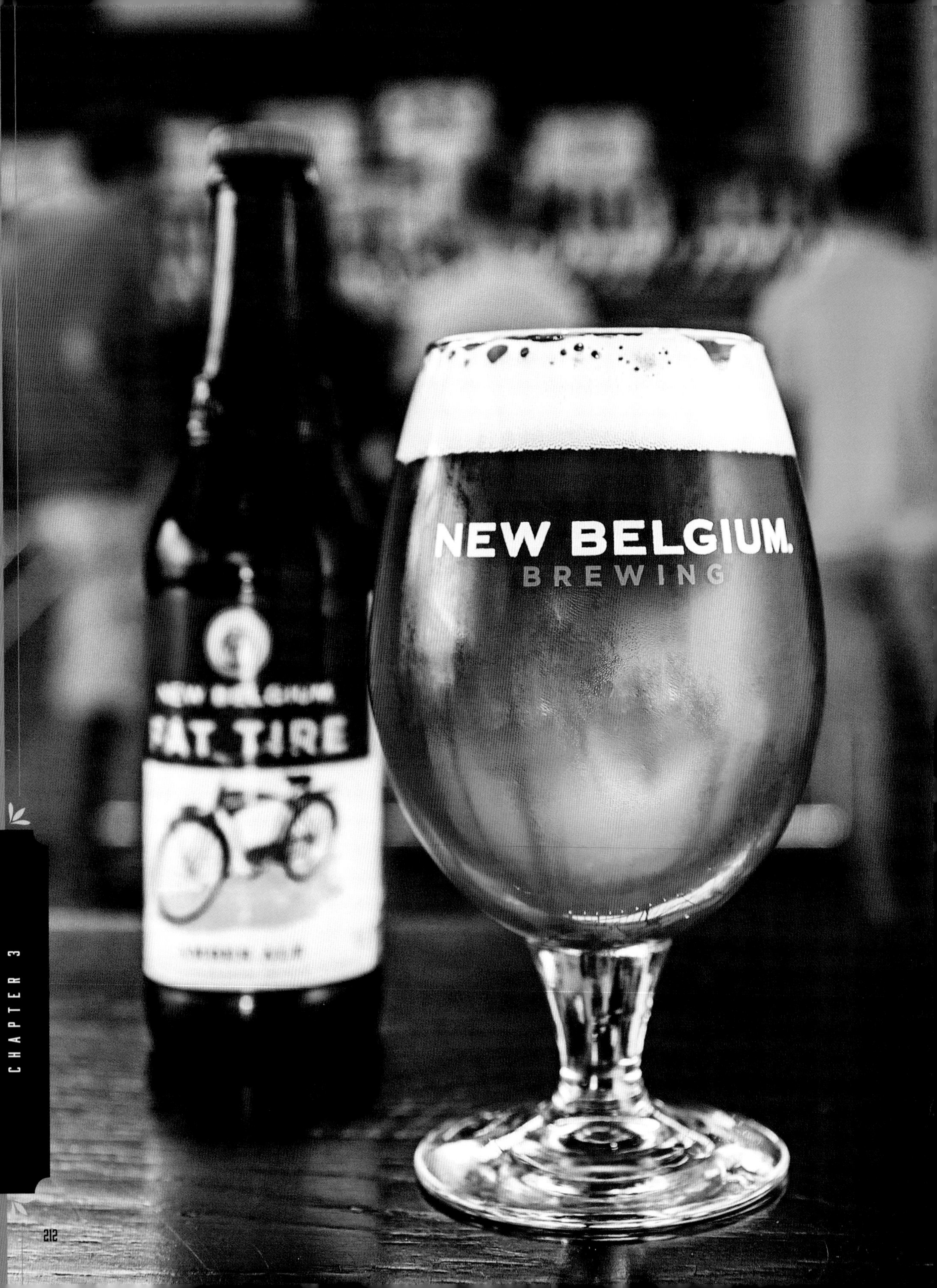
NEW BELGIUM.
BREWING
FAT TIRE

PAULANER BRAUEREI
PAULANER HEFE-WEIZEN

(5 gallons/19 L, all-grain) OG = 1.053 FG = 1.010 IBU = 18 SRM = 5 ABV = 5.6%

This beer from Paulaner is a well-balanced example of a hefeweizen. Watch your fermentation temperature to get the much sought after "breadiness" and banana/clove aroma. *Prost!*

INGREDIENTS

7.5 lb. (3.4 kg) wheat malt
3.25 lb. (1.5 kg) Pilsner malt
4.75 AAU Hallertau-Hersbrücker hops (60 min.) (1.2 oz./34 g
  at 4% alpha acids)

Wyeast 3638 (Bavarian Wheat), White Labs WLP380
  (Hefeweizen IV), or LalBrew Munich yeast
$^3/_4$ cup (150 g) dextrose (for priming)

STEP BY STEP

Perform a single decoction mash with a 30-minute rest at 131°F (55°C) and a 45-minute rest at 153°F (67°C). Mash out, vorlauf, and then sparge at 170°F (77°C). Boil for 120 minutes, adding the hops as instructed. Pitch the yeast and ferment at 68°F (20°C). Rack to secondary when fermentation is complete. Bottle or keg a few days later, when the beer falls clear.

EXTRACT WITH GRAINS OPTION: Replace the wheat malt and Pilsner malt with 1.5 pounds (0.68 kg) Briess dried wheat malt extract, 3.75 pounds (1.7 kg) Weyermann Bavarian Hefeweizen liquid wheat malt extract (late addition), 2.1 pounds (0.95 kg) wheat malt, and 0.91 pounds (0.41 kg) Pilsner malt. In a large soup pot, heat 4.5 quarts (4.3 L) of water to 169°F (76°C). Add the crushed grains to your grain bag. Submerge the bag and let the grains steep around 158°F (70°C) for 45 minutes. While the grains steep, begin heating 2.1 gallons (7.9 L) of water in your brewpot. When the steep is over, remove 1.5 quarts (1.4 L) of water from the brewpot and add to the "grain tea" in steeping pot. Place a colander over the brewpot and place the steeping bag in it. Pour the diluted grain tea through the grain bag. Heat the liquid in the brewpot to a boil, then stir in the dried malt extract, add the first charge of hops, and begin the 60-minute boil. With 15 minutes left in the boil, turn off the heat and stir in the liquid malt extract. Stir well to dissolve the extract, then resume heating. Chill the wort, transfer to your fermenter, and top up with filtered water to 5 gallons (19 L). Follow the remaining portion of the all-grain recipe.

PEDAL HAUS BREWERY
BOURBON BARREL-AGED QUADRUPEL

(5 gallons/19 L, all-grain)   OG = 1.094   FG = 1.011   IBU = 18   SRM = 19   ABV = 10.6%

Pedal Haus Brewery is a star in Arizona's craft beer industry and their barrel-aged quadrupel is a beer big in all regards: aroma, flavor, mouthfeel, and alcohol. It's a complex blend with no one component standing out by itself.

INGREDIENTS

13 lb. (5.9 kg) 2-row pale malt
11 oz. (0.25 kg) dark Munich malt (30°L)
5.5 oz. (160 g) melanoidin malt
12.5 oz. (0.35 kg) Simpsons DRC malt
3 oz. (84 g) crystal malt (75°L)
1 oz. (28 g) acidulated malt
1.6 lb. (0.73 kg) dextrose (dextrose)
0.63 lb. (0.29 kg) amber candi syrup
0.63 lb. (0.29 kg) amber dried malt extract

3.3 AAU Magnum hops (90 min.) (0.25 oz./7 g
  at 13.1% alpha acids)
2 AAU Magnum hops (40 min.) (0.15 oz./4 g
  at 13.1% alpha acids)
1 oz. (28 g) American oak chips (Bourbon-soaked)
Wyeast 3787 (Belgian High Gravity), White Labs WLP530
  (Abbey Ale), or Mangrove Jack's M41 (Belgian Ale) yeast
$^7/_8$ cup (175 g) dextrose (if priming)

STEP BY STEP

In advance of brew day, submerge your oak chips in your favorite Bourbon for a minimum of 2 weeks. Using a thick mash of 1.25 quarts/pound (2.6 L/kg) of grain, mix grains (not the sugars) with 4.66 gallons (17.6 L) of water. Mash in, targeting an initial temperature at 112°F (44°C) for a beta glucan rest. Once the grain is added, add 1 teaspoon of gypsum and a ½ teaspoon of calcium chloride. Hold at this temperature for 20 minutes. Raise temperature to 149°F (65°C) and hold at this rest for 20 minutes. Raise temperature to 158°F (70°C) and hold for 30 minutes. Once mash is complete, mash out for 10 minutes at 170°F (77°C). Recirculate until clear, and then begin collecting wort. With the goal of achieving 5.5 gallons (20.8 L) of wort after boil (which will yield a 5-gallon/19-L batch), collect 7.9 gallons (29.8 L) of wort since this beer requires a long boil. The 120-minute boil is necessary to develop the depth of malt flavor while also helping with color development. Add the malt extract and the first hops addition at the 90-minute mark, then the second hop addition at the 40-minute mark. Add the sugar and syrup in the last 5 minutes of the boil. Whirlpool vigorously and let settle for 20 minutes before chilling.

Chill the wort to 66°F (19°C). Pitch your yeast starter and oxygenate thoroughly. Primary fermentation should take 5 days. Upon completion of primary, raise the temperature to 72°F (22°C) and add your Bourbon-soaked oak chips to replicate barrel aging. Hold at this temperature for 21 more days to fully develop the complex flavors of the beer. (Note: If fermentation stalls well short of the goal, it's OK to pitch a second helping of neutral yeast, such as SafAle US-05, WLP001, or Wyeast 1056. Just do not oxygenate during the second pitching. If you do a second yeast pitch, give that 5 days to finish primary, then begin the secondary fermentation process mentioned above.) When fermentation is complete, bottle or keg and carbonate to 2.8 v/v.

PARTIAL MASH OPTION: In advance of brew day, submerge your oak chips in your favorite bourbon for a minimum of 2 weeks. Reduce the 2-row pale malt to 1 pound (0.45 kg) and substitute with 7.5 pounds (3.4 kg) pale ale dried malt extract. Bring 2 gallons (7.6 L) up to about 160°F (70°C). Place the pale, Munich, and melanoidon malts in a muslin bag and steep for about 30 minutes. Add the crystal malts to a separate muslin bag and steep in the wort for an additional 15 minutes (45 minutes total). After, place all the grains in a colander and wash with 2 gallons (7.6 L) of hot water. Raise the temperature to near-boiling, then remove from heat and add the two malt extracts while stirring vigorously. Return to a boil and boil for 60 minutes, adding the hops as indicated, and add the sugar and syrup in the last 5 minutes of the boil. Once the boil is complete, top up with water with the goal of achieving 5.5 gallons (20.8 L) of wort, which will yield a 5-gallon (19 L) batch after fermentation is complete. Follow the remaining all-grain instructions for fermentation and packaging.

TIPS FOR SUCCESS

For the malts, brewmaster Derek "Doc" Osborne uses Simpsons DRC, which is very similar to Special B malt. Substitute if the Simpsons DRC is unavailable. If you cannot find Munich 30°L, using dark Munich at 20°L will suffice. Crystal 75 can be exchanged for Crystal 77 or Crystal 80, depending on what is available to you.

Anytime you are brewing beers >10% ABV, fermentation control is of the utmost importance. The yeast has to work really hard to ferment this much sugar, and temperatures can rise quickly, producing alcohols with higher molecular weight that can leave off-flavors in your beer. It's vital to pitch enough yeast and oxygenate thoroughly to ensure the yeast can do its job. Also oxygenate for a couple minutes. Doc says it's okay to add an additional burst of oxygen on day two as long as alcohol production has not yet started. This second oxygen addition can give the boost it needs to tackle such a big beer.

PELICAN BREWING COMPANY
KIWANDA CREAM ALE

(5 gallons/19 L, all-grain)   OG = 1.049   FG = 1.007   IBU = 25   SRM = 4   ABV = 5.4%

Not all late-hopped beers need to be aggressive. Pelican Brewing Co. from Pacific City, Oregon, created the first modern beer recipe we had heard about in which the first hopping addition was after flameout. This is the perfect summer beer to enjoy on the water.

INGREDIENTS

9 lb. (4.1 kg) 2-row pale malt
13 oz. (370 g) dextrin malt
8 oz. (230 g) flaked barley
10 AAU Mt Hood hops (0 min.) (2 oz./57 g
 at 5% alpha acids)

Wyeast 1056 (American Ale), White Labs WLP001
 (California Ale), or SafAle US-05 yeast
¾ cup (150 g) dextrose (if priming)

STEP BY STEP

Mill the grains and mix with 15.5 quarts (14.6 L) of 162°F (72°C) strike water to reach a mash temperature of 150°F (66°C). Hold this temperature for 90 minutes. Vorlauf until your runnings are clear. Sparge with enough water to obtain 6.5 gallons (24.6 L) of wort. Boil for 60 minutes. After the boil, turn off heat and begin a whirlpool of the hot wort. Add the hops while whirlpooling. Let stand for 20 minutes, then chill the wort to 65°F (18°C). Aerate with pure oxygen or filtered air and pitch yeast. Ferment at 65°F (18°C) for 7 days. Raise to 70°F (21°C) for 3 more days. Once the beer reaches terminal gravity, bottle or keg and carbonate to approximately 2.5 volumes. You can cold-crash the beer prior to packaging to 35°F (2°C) for 48 hours to improve clarity.

PARTIAL MASH OPTION: Reduce the pale malt to 1 pound (0.45 kg) and add in 4.25 pounds (2 kg) extra light dried malt extract. Bring 1 gallon (4 L) of water to approximately 160°F (71°C). Submerge the milled grains in grain bags and steep for 45 minutes, trying to maintain the temperature at about 150°F (66°C). Remove the grain bags, and place in a colander. Wash the grains with 1 gallon (4 L) of hot water. Add the dried malt extract while stirring, and stir until dissolved, then top off to 5 gallons (19 L). Boil for 30 minutes. After the boil, turn off heat and begin a whirlpool of the hot wort. Add the hops while whirlpooling. Let stand for 20 minutes, then chill the wort to 65°F (18°C) and transfer wort to the fermenter. Top off the fermenter to 5 gallons (19 L), then aerate with pure oxygen or filtered air and pitch yeast. Ferment at 65°F (18°C) for 7 days. Raise to 70°F (21°C) for 3 more days. Once the beer reaches terminal gravity, bottle or keg and carbonate to approximately 2.5 volumes. Cold-crash prior to packaging to 35°F (2°C) for 48 hours to improve clarity.

PLZENSKY PRAZDROJ'S
PILSNER URQUELL CLONE

(5 gallons/19 L, all-grain) OG = 1.048 FG = 1.015 IBU = 40 SRM = 4 ABV = 4.4%

Brewed in Plzen, Czech Republic, Pilsner Urquell is the original Pilsner beer. Brew this clone with soft water.

INGREDIENTS

8 lb. (3.6 kg) continental Pilsner malt
1 lb. (0.45 kg) Vienna malt
0.5 lb. (0.23 kg) Munich malt (6°L)
0.5 lb. (0.23 kg) Carafoam malt
5.2 AAU Saaz hops (80 min.) (1.3 oz./38 g
  at 4% alpha acids)
3.2 AAU Saaz hops (45 min.) (0.8 oz./23 g
  at 4% alpha acids)

3 AAU Saaz hops (25 min.) (0.75 oz./21 g
  at 4% alpha acids)
1 tsp. Irish moss
Wyeast 2001 (Pilsner Urquell H-strain), White Labs WLP800
  (Pilsner Lager), or Mangrove Jack's M84 (Bohemian Lager)
  yeast
³/₄ cup (150 g) dextrose (if priming)

STEP BY STEP

The traditional Pilsner Urquell utilizes an undermodified Pilsner malt that is malted by the brewery. The mash is subsequently triple decocted to help break down the protein matrix that is still present in the undermodified malts. For three reasons we have decided to change up the classic recipe. First, homebrewers don't often have access to undermodified Pilsner malts these days. Second, a triple decoction means a long and rigorous brew day for the brewer. Finally, adding in a protein rest could be detrimental to the beer's body and head retention. So we've simplified the mash procedure and added some malts to help adjust the color and add a slight toasted character that the triple decoction should otherwise add to the beer's profile.

Mash grains with 15 quarts (14.2 L) of soft water at 142°F (61°C) and hold for 30 minutes. Perform a decoction, pull ¹/₃ thick portion of the mash (~1.5 gallons/5.7 L), and bring to a boil. Boil for 15 minutes being sure not to scorch the grain mix. Stir the decocted portion back into the main mash. The main mash should now settle at 155°F (68°C). Hold for a 30-minute rest. Mash out, vorlauf, and then sparge at 170°F (77°C) to collect enough wort to result in 5 gallons (19 L) after a 90-minute boil. Boil for 90 minutes, adding hops at times indicated. Cool, aerate, and pitch yeast. Ferment at 50°F (10°C) until signs of fermentation have slowed considerably. Some diacetyl is noted in Pilsner Urqell, but if there is a strong diacetyl presence at this point, 2–4 days at 65°F (18°C) is advised. Rack to secondary and lager for 4–6 weeks at 40°F (4°C). After lagering is complete, bottle or keg as usual.

PARTIAL MASH OPTION: Replace grains in all-grain recipe with 1.4 pounds (0.64 kg) Pilsen dried malt extract, 3.3 pounds (1.5 kg) Pilsen light liquid malt extract, 1.75 pounds (0.79 kg) Pilsner malt, 0.5 pound (0.23 kg) Vienna malt, 0.25 pound Munich malt (10°L), and 0.5 pound (0.23 kg) Carapils malt. In a large soup pot, heat 4.5 quarts (4.3 L) of water to 169°F (76°C). Add crushed grains to grain bag. Submerge bag and let grains steep around 158°F (70°C) for 45 minutes. While grains steep, begin heating 2.1 gallons (7.8 L) of water in your brewpot. When steep is over, remove 1.5 quarts (1.4 L) of water from brewpot and add to the "grain tea" in steeping pot. Place colander over brewpot and place steeping bag in it. Pour grain tea (with water added) through grain bag. This will strain out any solid bits of grain and rinse some sugar from the grains. Heat liquid in brewpot to a boil, then stir in dried malt extract. Boil for 60 minutes, adding liquid malt extract with 15 minutes remaining and hops as indicated. Cool and top off to 5 gallons (19 L). Follow the remaining portion of the all-grain recipe.

REVOLUTION BREWING
LOUIE LOUIE

(5 gallons/19 L, all-grain) OG = 1.072 FG = 1.010 IBU = 58 SRM = 6 ABV = 8.2%

Revolution Brewing in Chicago, Illinois, decided to brew up a supposed "Zero IBU" IPA on their brewpub system. Using a whopping 5.5 pounds (2.5 kg) of hops per barrel, it was a hop bomb, but it should not technically have any "IBUs" based on isomerized alpha acids. What they found out was that according to lab analysis, there were still 58 IBUs in solution of the finished beer, derived from other compounds.

INGREDIENTS

12.9 lb. (5.85 kg) 2-row pale malt
1.25 lb. (567 g) red wheat malt
0.9 lb. (408 g) light Munich malt (6°L)
2.5 oz. (71 g) Amarillo hops (hopstand)
2.5 oz. (71 g) Centennial hops (hopstand)
2.5 oz. (71 g) Amarillo hops (1st round – dry hop)

2.5 oz. (71 g) Centennial hops (1st round – dry hop)
5 oz. (142 g) Mosaic hops (2nd round – dry hop)
Wyeast 1028 (London Ale), White Labs WLP013 (London Ale), or LalBrew Nottingham yeast
3/4 cup (150 g) dextrose (if priming)

STEP BY STEP

This recipe is designed to achieve 5.5 gallons (21 L) of wort in the fermenter on brew day. This will help offset the loss of volume to the heavy hopping rate of this beer. Mill the grains and mix with 5 gallons (19 L) of soft water at 163°F (73°C) to reach a mash temperature of 150°F (66°C). Hold this temperature for 60 minutes. Vorlauf until your runnings are clear. Sparge with enough water to obtain 7 gallons (25 L) of wort. You may want to add a pinch of hops or an additive like Fermcap to help with foam control. After the boil, turn off heat and chill wort to 165°F (74°C), then begin a whirlpool of the wort and add the hopstand hops. Let settle for 30 minutes, then chill the wort to 66°F (19°C). The aim is to have 5.5 gallons (21 L) of wort in your fermenter.

TIPS FROM THE PROS

Revolution Brewing's brewer John Palos offered his thoughts on the traditional bittering charge of hops with these hop-forward beers and whether they are becoming obsolete. He replied, "They're definitely not a thing of the past. In fact, I feel that with growing development of (nonpellet) hop products, many of which are boil-only, we will see growing interest in boil additions. I do see increasingly a simplification of hop bills, which I think is a positive. Some of the best beers I've had are made with stunningly simple recipes. The days of the IPAs with six different malts (including 10 percent crystal) are in the past, and I think three-plus hop additions are also disappearing. However, that bittering charge is still invaluable in many hoppy beers. I feel like it imparts a more stable, tangible, chewy bitterness that will last longer in packaged beer."

Aerate and pitch yeast. Ferment at 68°F (20°C) for 7 days. Add the first round of dry hops and wait for 3 days. Carefully rack the beer off the first round of hops, hopefully in a closed transfer situation into a CO_2-purged receiving vessel. Bag the second round of hops and add to the fermenter. Wait 2–3 days, then remove. Carbonate to approximately 2.5 volumes. You can cold-crash the beer prior to packaging to 35°F (2°C) for 48 hours to improve clarity.

EXTRACT-ONLY OPTION: Replace the malts from the all-grain recipe with 7 pounds (3.2 kg) extra light dried malt extract, 1.1 pounds (0.5 kg) wheat dried malt extract, and 0.66 pound (300 g) Munich dried malt extract. This recipe is designed to achieve 5.5 gallons (21 L) of wort in the fermenter on brew day. This will help offset the loss of volume to the heavy hopping rate. Heat 5 gallons (19 L) of soft water to 180°F (82°C) and remove from heat. Stir in all the dried malt extract and return to a boil. Boil wort for 15 minutes, then turn off heat and chill wort to 165°F (74°C). Follow the remainder of the all-grain recipe.

RUSSIAN RIVER BREWING COMPANY
PLINY THE ELDER

(5 gallons/19 L, all-grain) OG = 1.074 FG = 1.014 IBU = 100+ SRM = 6 ABV = 8.3%

Originally brewed to be part of a first-ever "Double IPA festival" back in 2000, Pliny has become the standard by which many modern double IPAs are measured. As with any hop-heavy beer, Pliny is best enjoyed fresh, while the massive hop aroma is at its peak.

INGREDIENTS

12.8 lb. (5.8 kg) 2-row pale malt
0.28 lb. (0.13 kg) crystal malt (45°L)
0.86 lb. (0.39 kg) Carapils malt
1 lb. (0.45 kg) dextrose
42.9 AAU Warrior hops (90 min.) (2.75 oz./78 g at 15.6% alpha acids)
6.1 AAU Chinook hops (90 min.) (0.5 oz./14 g at 12.2% alpha acids)
14.3 AAU Columbus hops (45 min.) (1 oz./28 g at 14.3% alpha acids)
12 AAU Simcoe hops (30 min.) 1 oz./28 g at 12% alpha acids)

20.5 AAU Centennial hops (0 min.) (2.25 oz./64 g at 9.1% alpha acids)
12 AAU Simcoe hops (0 min.) (1 oz./28 g at 12% alpha acids)
3.25 oz. (92 g) Columbus hops (dry hop)
1.75 oz. (50 g) Centennial hops (dry hop)
1.75 oz. (50 g) Simcoe hops (dry hop)
1 tsp. Irish moss (15 min.)
White Labs WLP001 (California Ale), Wyeast 1056 (American Ale), or SafAle US-05 yeast
3/4 cup (150 g) dextrose (if priming)

STEP BY STEP

Mash the grains at 150–152°F (66–67°C). Hold this temperature for 60 minutes. Mash out, vourlaf, and sparge. Boil the wort for 90 minutes, adding hops at the time indicated in the ingredients list. Chill the wort and pitch the yeast. Ferment at 68°F (20°C). Dry hop 2 weeks after primary fermentation slows for 5 days. Bottle or keg as usual.

EXTRACT WITH GRAINS OPTION: Replace the 12.8 pounds (5.8 kg) 2-row pale malt with 6.7 pounds (3 kg) extra light dried malt extract. Steep the crushed grains in 1 gallon (3.8 L) of water at 151°F (66°C) for 30 minutes. Rinse the grains with 2 quarts (2 L) of 170°F (77°C) water. Top up the kettle to 5.5 gallons and stir in the dried malt extract. Follow the remaining portion of the all-grain recipe.

SAPWOOD CELLARS BREWERY
CHEATER X

(5 gallons/19 L, all-grain) OG = 1.079 FG = 1.026 IBU = 80+ SRM = 4.5 ABV = 7.1%

Co-owner Michael Tonsmeire states, "This is the culmination of what we learned during our first year of brewing hazy IPAs. It utilizes two of our favorite 'cheater' hops for dry hopping for an intense aroma of passion fruit, gooseberry, and all sorts of citrus."

INGREDIENTS

8.5 lb. (3.9 kg) Rahr Standard 2-row malt
4 lb. (1.8 kg) Weyermann Pilsner malt
1.5 lb. (0.68 kg) Crisp naked malted oats
1.5 lb. (0.68 kg) Rahr white wheat malt
0.75 lb. (0.34 kg) BestMalz chit malt
1 oz. (28 g) Comet hops (mash)
3 oz. (85 g) Simcoe hops (0 min.)

2.5 oz. (71 g) Galaxy hops (0 min.)
5.5 oz. (156 g) Nelson Sauvin hops (dry hop)
5.5 oz. (156 g) Galaxy hops (dry hop)
RVA 132 (Manchester Ale), Wyeast 1318 (London Ale III),
  Imperial Yeast A38 (Juice), or LalBrew New England yeast
2/3 cup (130 g) dextrose (if priming)

STEP BY STEP

Mash at 156°F (69°C) with the mash hops added at the outset. Adjust brewing water by adding calcium chloride to achieve 150 ppm chloride and gypsum to achieve 150 ppm sulfate. If needed, add phosphoric acid to achieve a mash pH of 5.2. Collect wort and boil for 60 minutes, adding hops as noted. When the boil is complete, add the whirlpool hops without cooling the wort and allow to sit for 45 minutes before force chilling to 66°F (19°C). Transfer the chilled wort to an oversized fermenter. Aerate the wort well as this yeast is sensitive to underoxygenating, then pitch the yeast. Ferment at 68°F (20°C). Once the gravity stabilizes, chill to 55°F (13°C). Transfer to a purged keg with dry hops placed in screens. Dry hop for 3 days at 55°F (13°C), agitating once or twice daily for 30 seconds. Transfer off the hops to a serving keg. Pressurize to reach 2.4 volumes of CO_2.

PARTIAL MASH OPTION: Replace 2-row, Pilsner, and wheat malts with 4 pounds (1.8 kg) extra light dried malt extract, 2 pounds (0.91 kg) Pilsen dried malt extract, and 1.6 pounds (0.73 kg) wheat dried malt extract. In a muslin bag, heat the crushed oats, chit malt, and mash hops in 4 quarts (4 L) of water to 164°F (73°C). Mash at around 156°F (69°C) for 45–60 minutes. When the mash is done, remove the grains and wash with 1 gallon (4 L) of hot water. Bring volume up to 6 gallons (23 L) and stir in the malt extract. Follow the remainder of the all-grain recipe.

SIERRA NEVADA BREWING COMPANY
CELEBRATION

(5 gallons/19 L, all-grain) OG = 1.064 FG = 1.016 IBU = 65 SRM = 12 ABV = 6.8%

First brewed in 1981, Sierra Nevada explains that Celebration is one of the earliest examples of an American-style IPA, and it's still one of the few hop-forward holiday beers. The intense, hop-heavy beer features Chinook, Centennial, and Cascade hops.

INGREDIENTS

12.5 lb. (5.7 kg) 2-row pale malt
15 oz. (0.43 kg) caramel malt (60°L)
9 AAU Chinook hops (100 min.) (0.75 oz./21 g
  at 12% alpha acids)
5 AAU Centennial hops (100 min.) (0.5 oz./14 g
  at 10% alpha acids)
7.5 AAU Cascade hops (10 min.) (1.5 oz./43 g
  at 5% alpha acids)

0.66 oz. (19 g) Centennial hops (0 min.)
1.33 oz. (38 g) Cascade hops (0 min.)
1.33 oz. (38 g) Cascade hops (dry hop)
0.66 oz. (19 g) Centennial hops (dry hop)
Wyeast 1056 (American Ale), White Labs WLP001
  (California Ale), or SafAle US-05 yeast
1 cup (200 g) dextrose (for priming)

STEP BY STEP

Mash at 158°F (70°C) in 17 quarts (16 L) of water. Hold at this temperature for 60 minutes. Mash out, vorlauf, and then sparge at 170°F (77°C). Boil the wort for 100 minutes, adding the hops at times indicated. Pitch the yeast and ferment at 68°F (20°C). Dry hop in secondary for 5 days. Bottle or keg as usual.

PARTIAL MASH OPTION: Replace the 12.5 pounds (5.7 kg) 2-row pale malt in the all-grain recipe with 1 pound 1 ounce (0.48 kg) 2-row pale malt, 2.5 pounds (1.13 kg) light dried malt extract, and 5 pounds (2.27 kg) light liquid malt extract. Steep the grains at 157.5°F (69.7°C) in 3 quarts (2.9 L) of water. Rinse the grains with 2 quarts (2 L) of 170°F (77°C) of water. Add water to the brewpot to make at least 3 gallons (11 L) of wort. Stir in the dried malt extract and boil the wort for 100 minutes, adding hops at times indicated. Keep some boiling water handy and do not let the boil volume dip below 3 gallons (11 L). Add the liquid malt extract in the final 15 minutes of the boil. Stir thoroughly to avoid scorching. Chill the wort, transfer to your fermenter, and top up with filtered water to 5 gallons (19 L). Follow the remaining portion of the all-grain recipe.

SIERRA NEVADA BREWING COMPANY
PALE ALE

(5 gallons/19 L, all-grain) OG = 1.052 FG = 1.011 IBU = 38 SRM = 10 ABV = 5.4%

This signature pale ale was originally dreamed up as a homebrew. Now, decades later, it has launched thousands of homebrews in its wake. It's a crisp, hoppy classic by which all American pale ales are measured.

INGREDIENTS

10 lb. 2 oz. (4.6 kg) 2-row pale malt
11 oz. (0.3 kg) caramel malt (60°L)
4.4 AAU Perle hops (90 min.) (0.5 oz./14 g
   of 8.8% alpha acids)
6.0 AAU Cascade hops (45 min.) (1.0 oz./28 g
   of 6% alpha acids)

1.5 oz. (43 g) Cascade hops (0 min.)
Wyeast 1056 (American Ale), White Labs WLP001
   (California Ale), or SafAle US-05 yeast
   (1 qt./1 L yeast starter)
1 cup (200 g) dextrose (for priming)

STEP BY STEP

Two or three days before brew day, make a yeast starter, aerating the wort thoroughly (preferably with oxygen) before pitching the yeast. Mash the grains at 155°F (68°C). Mash out, vorlauf, and then sparge at 170°F (77°C). Collect 6.5 gallons (25 L) of wort. (Check that the final runnings do not drop below SG 1.010.) Boil for 90 minutes, adding the hops at the times indicated. Pitch the yeast and ferment at 68°F (20°C) until final gravity is reached. Bottle or keg as usual.

EXTRACT WITH GRAINS OPTION: Scale the 2-row pale malt down to 1 pound 5 ounces (0.60 kg) and add 1.75 pounds (0.80 kg) light dried malt extract and 4 pounds (0.79 kg) light liquid malt extract. Place the crushed grains in a steeping bag. Steep the grains at 155°F (68°C) in 3 quarts (2.9 L) of water. Remove the bag and place in a colander over the brewpot. Rinse the grains with 2 quarts (2 L) of 170°F (77°C) water. Add water to the brewpot to make at least 3 gallons (11 L) of wort. Stir in the dried malt extract and boil the wort for 90 minutes, adding the hops at the times indicated. Keep some boiling water handy and do not let the boil volume dip below 3 gallons (11 L). Add the liquid malt extract in the final 15 minutes of the boil. Chill the wort and transfer to your fermenter. Top up the fermenter to 5 gallons (19 L). Follow the remaining portion of the all-grain recipe.

TIPS FROM THE PROS

Be sure to pitch an adequate amount of yeast. The yeast starter size should allow you to yield the correct amount of yeast cells for a healthy fermentation. Aerate the starter well–preferably with oxygen–before pitching your yeast to the starter wort. If you aerate by shaking the starter, multiply the size of each starter by 1.33.

A little sulfate in your water will accentuate the hop character of the beer. You can add sulfate ions by adding gypsum to your brewing water. All-grain brewers starting with RO or distilled water should add 2 to 4 teaspoons per 10 gallons (38 L) of brewing water. Extract brewers can add 1 teaspoon of gypsum to the boil.

SOUTHERN TIER BREWING COMPANY
CRÈME BRÛLÉE

(5 gallons/19 L, all-grain) OG = 1.100 FG = 1.032 IBU = 55 SRM = 55 ABV = 10%

While many imperial stouts are all about re-creating chocolate desserts, Southern Tier had another idea: Why not crème brûlée? This inventive brew brings all the flavor of the classic hard-coated custard dessert into a creamy, dreamy beer.

INGREDIENTS

15.25 lb. (6.9 kg) 2-row pale malt
1.5 lb. (0.68 kg) flaked barley
1 lb. (0.45 kg) crystal malt (60°L)
1.5 lb. (0.68 kg) Belgian debittered black malt (600°L)
10 oz. (0.28 kg) lactose sugar (0 min.)
12 oz. (0.34 kg) caramelized white cane sugar (0 min.)
10.8 AAU Columbus hops (60 min.) (0.75 oz./21 g at 14.5% alpha acids)

9.2 AAU Chinook hop pellets (30 min.) (0.75 oz./21 g at 12.3% alpha acids)
3 vanilla beans split and deseeded (0 min.)
1 tsp. ground cardamom powder (0 min.)
$^1/_2$ tsp. yeast nutrient (15 min.)
$^1/_2$ tsp. Irish moss (15 min.)
White Labs WLP007 (Dry English Ale), Wyeast 1028 (London Ale), or LalBrew Nottingham yeast
$^2/_3$ cup (133 g) dextrose (if priming)

STEP BY STEP

This is a single-step infusion mash. Mix the crushed grains with 5.5 gallons (21 L) of 168°F (76°C) water to stabilize at 155°F (68°C) for 60 minutes. Sparge slowly with 175°F (79°C) water. Collect approximately 6.5 gallons (25 L) of wort runoff to boil for 60 minutes. During the boil you will want to make the caramelized sugar. Mix 12 ounces (0.34 kg) sugar in ¾ cup water in a saucepan over medium heat. Stir constantly until it turns to a thick liquid and becomes a medium amber color. Add to boiling wort immediately before it hardens. Make the other kettle additions as per the schedule. At the end of the boil let the wort rest 20 minutes and remove the vanilla beans. Cool the wort to 75°F (24°C). Pitch your yeast and aerate the wort heavily. Ferment at 68°F (20°C). After fermentation is complete, transfer to a carboy, and condition for 1 week before you bottle or keg.

EXTRACT WITH GRAINS OPTION: Reduce the 2-row pale malt in the all-grain recipe to 1.5 pounds (0.68 kg) and add 9 pounds (4.1 kg) light liquid malt extract. Steep the crushed grains in 2 gallons (7.6 L) of water at 155°F (68°C) for 30 minutes. Remove grains from the wort and rinse with 2 quarts (1.8 L) of hot water. Add the liquid malt extract and boil for 60 minutes. Follow the remaining portion of the all-grain recipe.

SURLY BREWING COMPANY
BENDER

(5 gallons/19 L, all-grain) OG = 1.057 FG = 1.013 IBU = 43 SRM = 27 ABV = 6%

Bender is a category-bending American brown ale brewed with oat crystal malt. It is described by the brewery as "crisp and lightly hoppy, complemented by the velvety sleekness oats deliver. Belgian and British malts usher in cascades of cocoa, bitter-coffee, caramel, and hints of vanilla and cream."

INGREDIENTS

7.5 lb. (3.4 kg) British 2-row pale malt
2 lb. (0.9 kg) Belgian aromatic malt
0.75 lb. (0.34 kg) British medium crystal malt (55°L)
0.75 lb. (0.34 kg) Belgian Special B malt (135°L)
0.63 lb. (0.28 kg) Simpsons Golden Naked Oats (10°L)
4 oz. (113 g) British chocolate malt (425°L)

1.25 AAU Willamette hops (first wort hop) (0.25 oz./7 g at 5% alpha acids)
10.5 AAU Columbus hops (60 min.) (0.75 oz./21 g at 13% alpha acids)
2.5 oz. (71 g) Willamette hops (0 min.)
Wyeast 1335 (British Ale II), White Labs WLP022 (Essex Ale), or Mangrove Jack's M15 (Empire Ale) yeast
³/₄ cup (150 g) dextrose (if priming)

STEP BY STEP

Mash the grains at 152°F (67°C) and hold for 60 minutes. Vorlauf until your runnings are clear and add the Willamette first wort hops to the kettle. Sparge the grains with 4 gallons (15 L) of water and top up as necessary to obtain 6 gallons (23 L) of wort. Boil the wort for 60 minutes, adding the Columbus hops at the beginning of the boil and the second Willamette addition at the end. After the boil, turn off the heat and chill the wort rapidly to 65°F (18°C). Aerate the wort with pure oxygen or filtered air and pitch the yeast. Ferment at 67°F (19°C) for 7 days. Increase the temperature to 72°F (22°C) for an additional 3 days. Bottle or keg the beer and carbonate to approximately 2 volumes.

PARTIAL MASH OPTION: Scale the British 2-row pale malt down to 1 pound (0.45 kg) and add 4.5 pounds (2 kg) golden light liquid malt extract. Bring 2 gallons (7.6 L) of water to approximately 165°F (74°C) to stabilize the mash at 152°F (67°C). Add the milled grains in grain bags to the brewpot to mash for 60 minutes. Remove the grain bags and wash the grains with 1 gallon (4 L) of hot water. Top off to 5.5 gallons (20 L) with water. Remove the brewpot from the heat and add the liquid extract while stirring. Stir until completely dissolved. Add the Willamette first wort hops addition, put your pot back on the heat, and bring the wort to a boil. Follow the remaining portion of the all-grain recipe.

TIPS FROM THE PROS

Bender relies to a significant degree on the use of British (note the British medium crystal) and Belgian (note the aromatic and Special B) malts to develop its unique malt characteristics. Golden Naked Oats is a huskless oat crystal malt from Simpsons that is added for a subtle, nutty flavor. It adds a smooth, oaty mouthfeel and a creamy head to your beer. Because the oats are a "crystal" malt, they do not need to be mashed.

TINY REBEL BREWING COMPANY
BITTER SWEET SYMPHONY

(5 gallons/19 L, all-grain) OG = 1.056   FG = 1.012   IBU = 30   SRM = 12   ABV = 5.7%

This Welch-based brewery brews up a nontraditional ESB, with a good portion of the hops coming from the United States. Feel free to substitute out the crystal 300 for another dark crystal malt with rich toffee or dark fruit character or use a darker Munich malt to compensate.

INGREDIENTS

8.5 lb. (3.86 kg) pale ale malt
1.25 lb. (0.57 kg) Munich malt
1.25 lb. (0.57 kg) torrified wheat
0.5 lb. (0.23 kg) crystal 300 malt (115°L)
2.8 AAU Cascade hops (60 min.) (0.35 oz./10 g
 at 8% alpha acids)

16 AAU Cascade hops (0 min.) (2 oz./57 g
 at 8% alpha acids)
1 oz. (28 g) Bramling Cross hops (dry hop)
2 oz. (57 g) El Dorado hops (dry hop)
Wyeast 1275 (Thames Valley), White Labs WLP023 (Burton
 Ale), or SafAle S-04 yeast
½ cup (100 g) dextrose (if priming)

STEP BY STEP

Mill the grains, then mix with 3.6 gallons (13.6 L) of 166°F (74°C) strike water to achieve a single-infusion rest temperature of 151°F (66°C). Hold at this temperature for 60 minutes. Vorlauf until your runnings are clear before directing them to your boil kettle. Batch or fly sparge the mash to obtain 6.5 gallons (25 L) of wort in the kettle. If fly sparing, stop once your running's SG hits 1.012 and top off to the desired preboil volume with water. Boil for 60 minutes, adding hops at the times indicated above left in the boil. At 15 minutes left in boil, add either Irish moss, Whirlfloc, or other kettle fining agent of your choice.

After the boil, add flameout hops and whirlpool for 15 minutes before rapidly chilling the wort to slightly below fermentation temperature, which is 70°F (21°C) for this beer. Pitch yeast and aerate well. Maintain fermentation temperature to avoid producing too many esters, which can easily occur with this strain. Once primary fermentation is done, drop the temperature to 61°F (16°C). Add the dry hops and let them extract for 5 days. Bottle or keg the beer and carbonate to approximately 2.0 volumes.

TIPS FOR SUCCESS

First off, let's talk about numbers. If you put the amount of hops into your favorite brewing software program, you'll likely find that the IBUs calculated are far from the value of 30 given. Fear not, the ratio of hops comes straight from Tiny Rebel's brewers and there will be plenty of hop character. Hop chemistry is a complex beast, and elements such as flavor, isomerization, etc., remain cloaked in mystery. As for water chemistry, Mark Gammons, Tiny Rebel's production manager, recommends a profile to target a traditional sweet ale flavor. Within these boundaries, shoot for 200 ppm total hardness with chlorides around 150 ppm. In addition, the pH of your wort prior to pitching should be around 5.1, which means that your mash pH should be at the upper end of the ideal mash pH range, 5.4–5.5.

PARTIAL MASH OPTION: Replace the pale ale and Munich malts with 4.5 pounds (2.04 kg) light dried malt extract and 1 pound (0.45 kg) Munich dried malt extract. Bring 6.5 gallons (25 L) of water to roughly 150°F (66°C). Steep the torrified wheat and crystal malt for 15 minutes before removing. Add the malt extracts, with stirring, before heating to a boil. Boil for 60 minutes, adding hops at the times indicated above left in the boil. At 15 minutes left in boil, add either Irish moss, Whirlfloc, or other kettle fining agent of your choice.

After the boil, add flameout hops and whirlpool for 15 minutes before rapidly chilling the wort to slightly below fermentation temperature, which is 70°F (21°C) for this beer. Pitch yeast. Maintain fermentation temperature to avoid producing too many esters, which can easily occur with this strain. Once primary fermentation is done, drop the temperature to 61°F (16°C). Add the dry hops and let them extract for 5 days. Bottle or keg the beer and carbonate to approximately 2.0 volumes.

TRAQUAIR HOUSE BREWERY
TRAQUAIR HOUSE ALE

(5 gallons/19 L, all-grain) OG = 1.075   FG = 1.019   IBU = 35   SRM = 13+   ABV = 7.2%

Traquair House Ale is a deep reddish, full-bodied, and richly flavored ale. It carries an alcoholic warmth, hop bitterness, and smoky malt flavor that is unmatched by any other beer.

INGREDIENTS

15 lb. (6.8 kg) 2-row pale malt
0.25 lb. (0.11 kg) roasted barley (300°L)
6 AAU Kent Goldings hops (90 min.) (1.5 oz./42 g at 4% alpha acids)

5 AAU Kent Goldings hops (30 min.) (1.25 oz./35 g at 4% alpha acids)
Wyeast 1728 (Scottish Ale) or SafAle S-04 yeast
³/₄ cup (150 g) dextrose (for priming)

STEP BY STEP

Mash the grains at 152°F (67°C) for 60 minutes. Put 1 gallon (3.8 L) of wort runoff into a brewpot and boil it for 30 minutes, stirring often. This makes a small amount of caramelized wort to be added later. Collect a further 7 gallons (26.6 L) of wort and begin the boil, which will last about 2 hours. When you have 6.5 gallons (24.7 L) of wort (and around 90 minutes left in the boil), add the caramelized wort and the first addition of Kent Goldings hops. Add the remaining Goldings with 30 minutes left in the boil. Pitch the yeast and ferment 8 to 10 days at 65°F (18°C), then transfer to secondary and condition at 50°F (10°C) for 2 weeks. Package and age in bottles or kegs for 8–10 weeks.

EXTRACT WITH GRAINS OPTION: Steep 2 pounds (0.91 kg) 2-row pale malt and 4 ounces (112 g) roasted barley in 3.4 quarts (about 3.4 L) of water. Steep at 152°F (67°C) for 45 minutes. Rinse the grains with 1.5 quarts (about 1.5 L) of 170°F (77°C) water. Add water to the "grain tea" to make 3 gallons (11 L), add 3 pounds (1.4 kg) light dried malt extract, and bring the wort to a boil. Boil for 60 minutes, adding 6.5 AAU of Kent Goldings hops at the beginning of the boil. Make other hop additions as specified in recipe. With 15 minutes left in the boil, turn off the heat and stir in 5.6 pounds (2.5 kg) of light liquid malt extract. Resume heating for the remainder of boil time. At the beginning of the main wort boil, scoop 2 quarts (about 2 L) wort into a 3-quart (about 3 L) secondary brewpot and boil alongside the main wort for 30 minutes. This will make some caramelized wort to add back later. Stir the "mini-wort" often to prevent scorching. After 30 minutes of reducing the "mini-wort," combine it with the main wort. Cool the wort, transfer to your fermenter, and top up to 5 gallons (19 L) with filtered water. Follow the remaining portion of the all-grain recipe.

TRILLIUM BREWING COMPANY
FORT POINT PALE ALE

(5 gallons/19 L, all-grain)   OG = 1.060   FG = 1.013   IBU = 45   SRM = 5   ABV = 6.6%

Trillium's website describes this beer as "layers of hops-derived aromas and flavors of citrus zest and tropical fruit rest on a pleasing malt backbone. Dangerously drinkable with a dry finish and soft mouthfeel from wheat. Our year round hoppy pale ale culminates in a restrained bitterness and dry finish."

INGREDIENTS

10 lb. (4.3 kg) 2-row pale malt
1.5 lb. (0.68 kg) wheat malt
12 oz. (0.34 kg) dextrin malt
4 oz. (113 g) British pale crystal malt (22°L)
3.5 AAU Columbus hops (60 min.) (0.25 oz./7 g at 14% alpha acids)
10.5 AAU Columbus hops (10 min.) (0.75 oz./21 g at 14% alpha acids)

2 oz. (57 g) Columbus hops (hop stand)
4 oz. (113 g) Citra hops (dry hop)
1 oz. (28 g) Columbus hops (dry hop)
½ Whirlfloc tablet (10 min.)
White Labs WLP007 (Dry English Ale), Wyeast 1098 (British Ale), Imperial Yeast A04 (Barbarian), or LalBrew New England yeast
¾ cup (150 g) dextrose (if priming)

STEP BY STEP

Crush the malt and add to 4 gallons (15 L) of strike water to achieve a stable mash temperature at 150°F (65.5°C) until enzymatic conversion is complete. Sparge slowly with 170°F (77°C) water, collecting wort until the preboil kettle volume is 6 gallons (23 L). Boil the wort for 60 minutes, adding the hops as indicated. After the boil is finished, cool the wort to 180°F (82°C) and then add the hop stand addition. Stir the wort, then let settle for 30 minutes before chilling the wort down to yeast-pitching temperature. Now transfer to the fermenter and pitch the yeast. Ferment at 68°F (20°C). As the kräusen begins to fall, typically day 4 or 5, add the dry hops to the fermenter and let the beer sit on the hops for 5 days. Bottle with priming sugar or keg and force carbonate to 2.4 volumes CO_2.

EXTRACT WITH GRAINS OPTION: Replace the 2-row pale malt and wheat malt in the all-grain recipe with 6.6 pounds (3 kg) golden liquid malt extract and 1 pound (0.45 kg) wheat dried malt extract. Reduce the dextrin malt to 4 ounces (113 g). Place the crushed malt in a muslin bag. Steep the grains in 1 gallon (4 L) of water at 160°F (71°C) for 20 minutes. Remove the grain bag and wash with 2 quarts (2 L) of hot water. Top off the kettle to 5 gallons (19 L) and heat up to boil. As soon as the water begins to boil, remove the brewpot from the heat and stir in the dried and liquid malt extracts. Stir until all the extract is dissolved, then return the wort to a boil. Boil the wort for 60 minutes. Follow the remaining portion of the all-grain recipe.

RESOURCES
BOOKS

American Sour Beers: Innovative Techniques for Mixed Fermentations, by Michael Tonsmeire (Georgetown, TX: Brewers Publications, 2017)

Former *BYO* "Advanced Brewing" columnist (2015–2018) Michael Tonsmeire provides inspiration, education, and practical applications for brewers of all levels with the help of some of the country's best-known sour beer brewers.

Brew Your Own's Big Book of Clone Recipes (Beverly, MA: Voyageur Press, 2018)

This book contains 300 commercial clone recipes, from the classics to the newest ales and lagers.

Brewing Better Beer: Master Lessons for Advanced Homebrewers, by Gordon Strong (Georgetown, TX: Brewers Publications, 2011)

Three-time Ninkasi Award–winning homebrewer and current *BYO* "Style Profile" columnist Gordon Strong digs deep into more advanced brewing techniques for those who want to take their brewing to the next level.

Brewing Classic Styles: 80 Winning Recipes Anyone Can Brew, by Jamil Zainasheff and John Palmer (Georgetown, TX: Brewers Publications, 2007)

Longtime *BYO* "Style Profile" columnist (2007 to 2013) Jamil Zainasheff has brewed every style of beer listed in the Beer Judge Certification Program guidelines. In this book, he teams up with *How to Brew* author John Palmer for a solid collection of recipes to get your classic beer styles dialed in.

The Complete Joy of Homebrewing, 4th edition, by Charlie Papazian (New York: William Morrow Paperbacks, 2014)

The classic title written by the founder of American homebrewing.

Designing Great Beers: The Ultimate Guide to Brewing Classic Beer Styles, by Ray Daniels (Georgetown, TX: Brewers Publications, 1998)

A practical, mechanical approach for when you are ready to start concocting your own homebrew recipes.

For the Love of Hops: The Practical Guide to Aroma, Bitterness and the Culture of Hops, by Stan Hieronymus (Georgetown, TX: Brewers Publications, 2012)

Part of the Brewing Elements series of titles from the Brewers Association, this reference is a must-have for any homebrewer interested in understanding hop varieties and hopping techniques.

How to Brew: Everything You Need to Know to Brew Right the First Time, by John Palmer (Georgetown, TX: Brewers Publication, 2006)

Many homebrewers learned to make beer thanks to John Palmer's comprehensive step-by-step guide. The original edition is also available online.

Malt: A Practical Guide from Field to Brewhouse, by John Mallett (Georgetown, TX: Brewers Publications, 2014)

Another title in the Brewers Association's Brewing Elements series. Bell's Brewery Director of Brewing Operations John Mallett will tell you everything you need to know about this backbone ingredient of beer.

Modern Homebrew Recipes: Exploring Styles and Contemporary Techniques, by Gordon Strong (Georgetown, TX: Brewers Publications, 2015)

Gordon Strong explains and explores the newest changes in Beer Judge Certification Program style guidelines and shares advice for recipe formulation, adapting recipes to your system, and many recipes.

Radical Brewing: Recipes, Tales, and World-Altering Meditations in a Glass, by Randy Mosher (Georgetown, TX: Brewers Publications, 2004)

Randy Mosher's style-busting guide to brewing great and artistic homebrews.

Simple Homebrewing: Great Beer, Less Work, More Fun, by Drew Beechum and Denny Conn (Georgetown, TX: Brewers Publications, 2019)

The third book from *BYO's* "Techniques" columnists, Drew Beechum and Denny Conn reduce the complicated steps for making beer and return to the fundamentals of extract and all-grain brewing, recipe design, and more.

Water: A Comprehensive Guide for Brewers, by John Palmer and Colin Kaminski (Georgetown, TX: Brewers Publications, 2013)

Another title in the Brewers Association's Brewing Elements series, this title takes the mystery out of one of the least understood but most important components of brewing.

Yeast: The Practical Guide to Beer Fermentation, by Chris White and Jamil Zainasheff (Georgetown, TX: Brewers Publications, 2010)

White Labs' founder Chris White teams up with Heretic Brewing Co.'s Jamil Zainasheff to explain yeast selection, storing and handling yeast, and much more about this crucial brewing ingredient. Also a part of the Brewers Association Brewing Elements series.

WEBSITES

Brew Your Own magazine: www.byo.com
The online home of *Brew Your Own*, featuring many years of *BYO* stories, resource guides, brewing calculator, charts, photo galleries, recipes, and more.

American Homebrewers Association: www.homebrewersassociation.org
Visit the AHA's site for lots of homebrew content, including joining the organization, events, competition guidelines, forums, recipes, and *Zymurgy* magazine.

How to Brew: www.howtobrew.com
Check out John Palmer's online version of his book, *How to Brew*.

Beer Judge Certification Program: www.bjcp.org
If you want to brew to style, this site is indispensable. It contains an online directory of the most up-to-date style guidelines, which are used in homebrewing competitions around the country (and the world).

Homebrew Academy (formerly known as Billy Brew): http://homebrewacademy.com
A collection of homebrewing articles and instructions by Billy Broas, all designed to teach people to make beer at home.

The Mad Fermentationist: www.themadfermentationist.com
A log of the homebrewing and fermentational experiments of The Mad Fermentationist—a.k.a. Michael Tonsmeire.

Brülosophy: www.brulosophy.com
Marshall Schott, Ray Found, Greg Foster, Malcolm Frazer, and Matt Waldron share a multitude of homebrewing experiments in their blog, plus lots of great info on brewing methods, recipes, and other projects.

Experimental Brewing: www.experimentalbrew.com
Homebrew buddies for life, Drew Beechum and Denny Conn, share a love for brewing the experimental beers. Check out their site for a blog, podcasts, books, speaking dates, and more.

Basic Brewing Radio: www.basicbrewing.com
James Spencer's podcast, which welcomes a wide variety of homebrewers on a multitude of topics. Features an archive of shows going back to 2005.

The Brewing Network: www.thebrewingnetwork.com
Home of *Beer Radio* podcasts and videos including *The Session*, *The Sour Hour*, *Dr. Homebrew*, *Brew Strong*, *Brewing with Style*, *Can You Brew It?*, *The Jamil Show*, and *The Home Brewed Chef*.

Beer and Wine Journal: http://beerandwinejournal.com
The current home of *BYO* former editor Chris Colby, created in conjunction with Basic Brewing Radio's James Spencer. It features many posts on all things homebrewing, from techniques, to ingredients, to styles, to recipes.

Larsbog: https://www.garshol.priv.no/
Lars Marius Garshol chronicles his research into farmhouse brewing in Europe, documenting what he learns through reading old archive collections, documents, and expeditions to remote parts of Europe.

TOOLS AND CALCULATORS

BeerSmith: www.beersmith.com
Home of Brad Smith's brewing software, but also features tons of information about beer styles, ingredients, recipes, a podcast, and a forum.

Brewer's Friend: www.brewersfriend.com
An online collection of homebrewing tools, including calculators, charts, water profiles, brew day sheets for note taking, water profiles, and much more. (Also available as an app for both iOS and Android.)

Mr. Malty: www.mrmalty.com
A collection of information and tools created by Jamil Zainasheff—former *BYO* writer, Ninkasi award winner, book author, podcast host, and current brewmaster/owner of Heretic Brewing Co.—back in his homebrewing days.

Bru'n Water: brunwater.com
Martin Brungard's water calculator used by brewers of all levels. Bru'n Water gives the brewer the ability to assess what minerals, acid, or dilutino should be incorporated into their brewing water.

MESSAGE BOARDS/FORUMS

HomeBrewTalk.com: www.homebrewtalk.com

Reddit /r/homebrewing: www.reddit.com/r/homebrewing

INDEX

ACKNOWLEDGMENTS

We would like to thank all of the wonderful BYO staff who have contributed to reviewing our content over the years, as well as the many fantastic freelance writers and editors who have made *Brew Your Own* great, including:

Fal Allen	*Ashton Lewis*	*Brad Smith*
Pattie Aron	*Dave Loew*	*Andy Sparks*
Steve Bader	*Dennis Maicupa*	*Jon Stika*
Glenn BurnSilver	*John Oliver*	*Gordon Strong*
Chris Colby	*John Palmer*	*Michael Tonsmeire*
Michael Dawson	*Steve Parkes*	*Joe Vella*
Terry Foster	*Betsy Parks*	*Josh Weikert*
Dave Green	*Sean Paxton*	*Chris White*
Aaron Hyde	*Bill Pierce*	*Chris Wood*
Christian Lavender	*Dan Russo*	*Jamil Zainasheff*

ABOUT THE AUTHOR

Brew Your Own, launched in 1995, is the largest-circulation magazine for people interested in making their own great beer at home. Every issue includes recipes, how-to projects, and expert advice to help you brew world-class beer. *Brew Your Own* publishes eight issues annually from offices in Manchester Center, Vermont. The magazine's online home, www.byo.com, offers a selection of the magazine's stories, projects, and recipes as well as web-only features. The magazine is available in both print and digital editions. Editor Dawson Raspuzzi joined *Brew Your Own* in 2013. He is a 2007 graduate of the journalism program at Castleton University in Castleton, Vermont. He lives with his patient wife, three rambunctious kids, and hound dog in Southern Vermont.